pretty little purses & pouches

pretty little purses & pouches

LARK BOOKS
A Division of Sterling Publishing Co., Inc.
New York / London

SENIOR EDITOR
Valerie Van Arsdale Shrader

EDITOR
Nathalie Mornu

ART DIRECTOR
Megan Kirby

COVER DESIGNERS
Dana Irwin
Cindy LaBreacht

ILLUSTRATIONS
Susan McBride

TEMPLATES
Orrin Lundgren

PHOTOGRAPHER
Stewart O'Shields

Library of Congress Cataloging-in-Publication Data

Pretty little purses & pouches / [Valerie Van Arsdale Shrader, editor].
-- 1st ed.
p. cm.
Includes index.
ISBN 978-1-60059-214-0 (hbk. : alk. paper)
1. Handbags. 2. Pouches (Containers) I. Shrader, Valerie Van Arsdale.
TT667.P73 2008
646.4'8--dc22
2008007973

10 9 8 7 6 5 4 3 2 1

First Edition

Published by Lark Books, A Division of Sterling Publishing Co., Inc.
387 Park Avenue South, New York, NY 10016

Distributed in Canada by Sterling Publishing,
c/o Canadian Manda Group, 165 Dufferin Street
Toronto, Ontario, Canada M6K 3H6

Distributed in the United Kingdom by GMC Distribution Services,
Castle Place, 166 High Street, Lewes, East Sussex, England BN7 1XU

Distributed in Australia by Capricorn Link (Australia) Pty Ltd.,
P.O. Box 704, Windsor, NSW 2756 Australia

If you have questions or comments about this book, please contact:
Lark Books, 67 Broadway, Asheville, NC 28801, 828-253-0467

Manufactured in China

ISBN 13: 978-1-60059-214-0

For information about custom editions, special sales, and premium and corporate purchases, please contact the Sterling Special Sales Department at 800-805-5489 or specialsales@sterlingpub.com.

contents

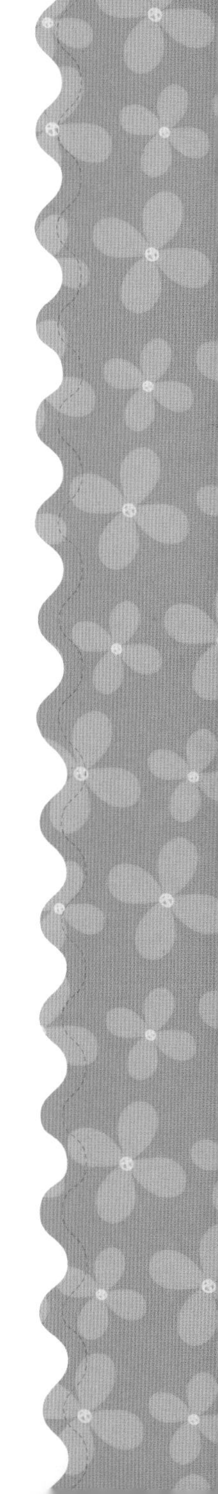

introduction

Remember playing dress up with your mother's purse? Or maybe your grandma let you rummage through her bag, and there, under the coin purse and the scarf, the lipsticks, the aspirin, and the old family photos—score!—sticks of gum. As kids, we learned that a woman's purse is a window into her world, and we couldn't wait to have one of our own.

The wait is over: The purse of your dreams is in these pages. *Pretty Little Purses & Pouches* gives you 29 reasons to never leave home without the perfect bag at your side. From pouches to wallets to shoulder bags, you can carry your life in exactly the style you want because *you* pick the pattern and the fabric. We put out a call for the coolest bags imaginable, and our designers delivered. They even stitched up some pretty little extras, like a jewelry roll and a phone cozy.

This book groups the projects by their dimensions, so you can size up the purse that fits your life. Need something just big enough to carry your cell phone, lip gloss, and a little bit of cash? Check out the first chapter, Just Right. Maybe you need some extra space and a stash pocket or two. That's where chapter two, Bring It, comes in. Or it could be you're the kind of gal who totes around everything but the kitchen sink. Well, then, Carry On to the last chapter, which contains the larger bags. No matter which way you turn, you'll find the perfect purse for every occasion.

For an elegant party, stitch up The Duchess (page 88) from flocked linen. Dotted Bliss (page 107), made from denim and polka-dotted fabric, is perfect for lazy weekend afternoons. If retro is your thing, you'll fall in love with the Dream On clutch on page 61. It's made from a vintage pillowcase and features a dreamy bow and zippered top. Give an old sweater you no longer wear a second life as a cool little tote by turning it into Eco Chic (page 70).

Make your purses even prettier with simple embellishments like yo-yos, patchwork, and buttons. You won't have to jump through hoops to embroider the simple image on Nest Egg (page 48). Need a little sparkle for an evening on the town? Deck yourself out with the pyramid-shaped Heavy Metals (page 76), which is spangled with sequins. Go with playful appliqué on Tokyo Rose (page 34).

Fashions come and fashions go, but everything in *Pretty Little Purses & Pouches* is made to stay. Looking for style? Flip the pages, grab some fabric, and get stitching. It's in the bag.

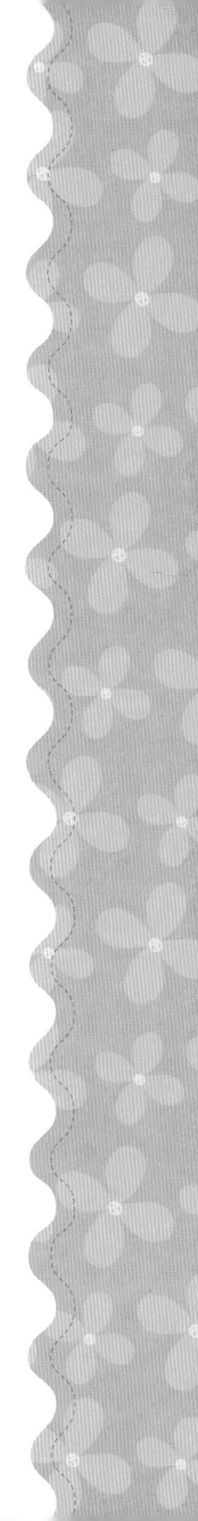

pretty little purses & pouches basics

This section is your foundation for the projects that follow. From the general supplies you'll need, to the specific tools, tricks, and techniques for that one special bag, you'll want to refer to it often.

tools

All of the projects use the supplies in the Basic Purses Tool Kit (at right) so start by hunting through your sewing basket, craft cupboard, and junk drawer to gather the items listed there. It's so much easier to have everything on hand than to go searching for something in the middle of a project, just when you've got your creative groove on.

Tool kit in hand, you'll want to read over the rest of this section to familiarize yourself with the other tools, materials, and techniques for these projects. Even a seasoned sewer may find important details and inspiration here.

SEWING SCISSORS

Sharp, quality scissors are a necessity for any sewing kit. Beware getting near paper with them—it would be shear madness! Paper's wood fibers will dull the blades quickly and make them useless on fabric. For basic fabric cuts, use a pair of 7- or 8-inch (17.8 or 20.3 cm) dressmaker's, bent-handled shears. These will allow the fabric to remain flat as you cut. For tight curves and other detail work, like trimming seams, use a pair of fine-tipped, 4- or 5-inch (10.2 or 12.7 cm) sewing scissors.

basic purses tool kit

- *Sharp sewing scissors (for fabric)*
- *Craft scissors (for paper)*
- *Measuring tape and transparent ruler*
- *Fabric marking pen and tailor's chalk*
- *Straight pins*
- *Scrap paper*
- *Pencil with an eraser*
- *Sewing machine*
- *Thread*
- *Sewing machine needles*
- *Seam ripper (nobody's perfect)*
- *Hand-sewing needles*
- *Needle threader*
- *Iron*

CRAFT SCISSORS

A cheap pair of craft scissors is all you need for these projects. Use them to cut out your paper templates. Find a pair that feels good in your hand, with a comfortable grip and moderate length so you can make expert cuts on curves and corners.

MEASURING TAPE & TRANSPARENT RULER

Essential for any sewing project, a measuring tape will keep you on the mark. A transparent ruler may be helpful, too, if you need to draw straight lines or take small measurements.

FABRIC MARKING PEN & TAILOR'S CHALK

Use a water-soluble fabric marker for marking lines to cut, embroider, or sew. The ink should vanish with plain water, but test your pen on a fabric scrap first, as the dyes in some fabrics can make the ink hard to remove. Fabric chalk is also helpful for quickly marking which cut pieces are fronts or backs, or left or right sides.

STRAIGHT PINS

Short metal pins with tiny heads will do the job for these projects, but longer ones with plastic or glass heads add a little color and fun. They're easier to handle and see too.

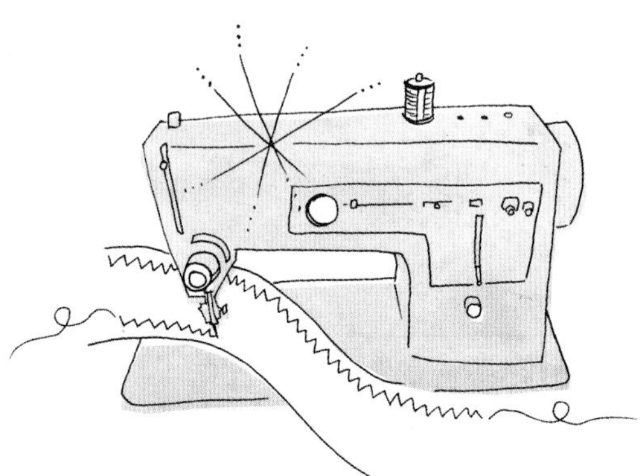

SEWING MACHINE

A few of the small purse projects in this book are stitched by hand, but most require a sewing machine. When machine sewing thicker fabrics, choose a longer stitch and reduce the pressure on the presser foot a bit to allow the material to move easily through the feed. Use a backstitch at the start and finish of any seam that needs a good anchor. Use a zigzag stitch on raw edges to keep fabrics from raveling.

SEWING MACHINE NEEDLES

Always have a stash of these on hand. One may break when you're sewing through many thicknesses of fabric, or it may become dull as you nick a pin here and there. They're inexpensive, and it's not worth damaging your fabric by trying to get by on an old dull one. Start each project with a new needle.

DON'T NEEDLE ME!

Regular woven fabrics—especially silk—do best with sharp needles. Save ballpoint needles (and pins) for knit materials. The ballpoint pushes between fibers, rather than piercing through and causing a run.

SEAM RIPPER

Quickly undo any stray or misguided stitching with this little treasure. It's also handy for removing basting stitches after their purpose is done.

HAND-SEWING NEEDLES

Some projects will have embellishments or a tight area that calls for hand stitching. A variety pack of needles in your kit will meet the need.

NEEDLE THREADER

This little tool consists of nothing more than a fine loop of wire attached to a small holder. Push the stiff, thin wire easily through the eye of any needle, insert a thread through the gaping wire loop, and pull the threader back out of the needle. The wire carries the thread back with it, threading the needle *sans* frustration.

IRON

Indispensable if you're working with cotton or linen, an iron is also needed to apply interfacings and appliqués, and to press seams open.

OPTIONAL TOOLS

Here's a sampling of additional tools you may need, depending on the project you select.

PINKING SHEARS

Pinking shears have a serrated edge; the resulting zigzag pattern can help limit fraying. Also use them to make decorative cuts and edges, as in the Spring Break clutch (page 83).

GLUE AND ADHESIVES

Spray adhesive, white glue, glue sticks, or a hot glue gun are all alternate ways to secure fabric or embellishments. The Patch Madame bag (page 104), for example, uses a glue stick as a quick alternative to basting.

ROTARY CUTTER AND MAT

Many of these projects could benefit from a rotary cutter. It's an easy way to cut lots of fabric pieces quickly. In general, the larger the rotary blade, the easier it is to cut fabric, but a midsize cutter will do the job for the projects in this book. Always use the rotary cutter with a mat. If you've never used a rotary cutter, try one at your local fabric store. You may want to roll with it!

EMBROIDERY NEEDLES

Some of these projects call specifically for an embroidery needle for detailing. This needle has a longer eye for ease in threading multiple strands of floss at once, but can be used for regular hand stitching as well.

EMBROIDERY HOOP

A hoop keeps the fabric taut while you embroider, so it's much easier to do the stitching. A small one will work for most of these projects, like the Happy Village pouch (page 38).

materials

THREADS

A quality polyester thread is all you need for your machine and hand sewing to create strong seams. A quality, all-cotton thread can also work well with the woven, natural fabrics (like cotton, for example) used in most of these projects. Whatever you do, stay away from bargain thread, which will undoubtedly break, fray, shed, knot, or otherwise wreck your project.

FLOSSES

Embroidery floss is a decorative thread that comes in six loosely twisted strands. It's available in cotton, silk, rayon, and other fibers, in a vast array of colors. Use multiple strands at once to sew decorative stitches.

PRESSING CLOTH

Have a lightweight piece of cotton—like a flour-sack kitchen towel—available to use as a pressing cloth for fabrics such as corduroy or velvet that require extra care when pressing on the right side. Otherwise, the iron may leave its mark on the nap.

INTERFACING

You can add support and structure to your projects with interfacing. The type used most often in the projects here is fusible interfacing, which means that you apply it by using the heat and pressure of an iron. Follow the manufacturer's instructions for fusing.

BIAS TAPE

Bias tape can be purchased as single fold or double fold in various widths. It's a handy edge binding because it stretches to fit a curved edge without puckering up the way a strip cut on the straight grain would. For a special project, make your own using the technique described on page 23.

fabrics

Purses need a material that doesn't give, something sturdy that will hold its shape whether you travel light or load it down. So these projects call for woven fabrics, not knits.

COTTON

Cotton is a natural choice because it's easy to sew and comes in a great array of colors and prints. Cotton's strengths are durability, density, and drape. A medium-weight cotton will do well in purse projects that use interfacing. Heavier cotton, even upholstery fabric, can be used for a sturdier, more durable purse. (If you want a purse that's washable, be sure to pre-shrink your cotton fabric.)

CANVAS

Canvas is a heavy-duty fabric used for sails, tents, and backpacks. It's a great choice for making a sturdy carryall purse. In the United States, canvas grades run in reverse of the weight, so number 10 canvas is lighter than number 4.

WOOL

Wool can be fuzzy or smooth, fleecy or ribbed. It's soft and strong and very absorbent. It's another great fabric for sewing.

LINEN

Linen is lustrous, strong, and tough. Like cotton, you can find it in an assortment of wonderful colors and prints. Linen is prone to wrinkling, so it's best to press your fabric both before and after sewing to have a smooth surface.

HOLEY HANDBAG, BATMAN

When repurposing old wool, be sure to clean it thoroughly. Any stains or perspiration on the fabric are a draw to moths, whose larvae will munch on your handbag.

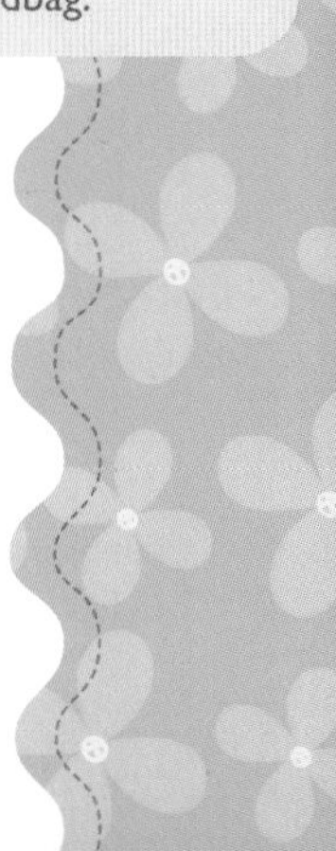

SILK

While linen is lustrous, silk is just plain luxurious, with a cost to match. Luckily, for most purses you need so little material that the project won't be pricey, and in many cases, you can even make use of scraps and remnants from other sewing projects. Choose a medium- to heavy-weight silk that can stand up to some wear and tear, and be sure to use a sharp, new needle when sewing. Silk also ravels or frays easily, so finish the seams inside your bag with this material. **Note:** Silk can be damaged by overexposure to sunlight and by oils.

USE WHAT YOU HAVE

If your purse project calls for batting or flannel to add dimension, you can substitute a number of materials you already have on hand: bleach-stained towels, old sweaters, worn out jeans, or a holey blanket. These will work just as well, and you can feel good about recycling them.

FELT

Felt is a wonder fabric. It's nonwoven, doesn't ravel, is soft and cuddly, and has no right side or wrong side. What could be easier to sew with? Although it's traditionally made from wool, synthetic felt is also available. Use either type for appliqué in these projects. For the body of a purse, though, real felted wool beats synthetic hands down.

BATTING AND LAYERING MATERIALS

Some of these purse projects call for batting or a layering material to add dimension. A common choice for this is cotton batting. It's sturdy, easy to work with, and available in different thicknesses. The Patch Madame bag (page 104) calls for batting, flannel, or muslin—depending on your preference for the weight and thickness of your bag.

embellishments

Add fancy little flourishes to your bags to make them sing.

RIBBON

Fabric ribbon is used in a number of these projects. From grosgrain to satin to velvet, ribbon adds dimension, texture, and flair to your bag.

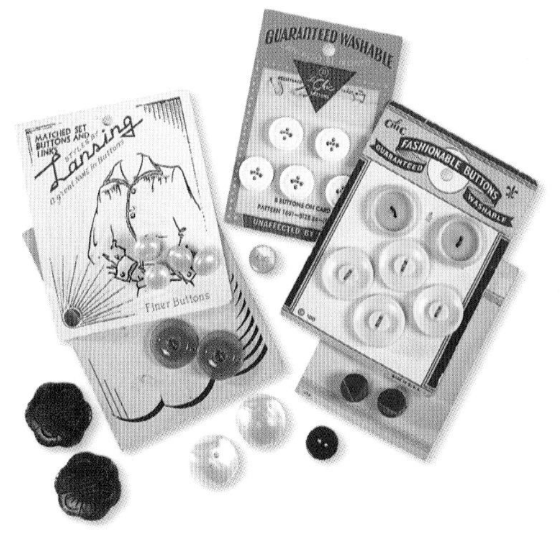

BEADS AND BUTTONS

The sky's the limit on bead and button options for decking out your purse. The Ruffled Delight bag (page 110) uses beads inside the straps for nicely weighted, good-to-grip handles. The Miss PR handbag (page 96) and Button Clutchin' (page 58), on the other hand, use buttons just for show. You may use commercial buttons, or purchase blanks to cover with fabric for an especially coordinated look.

COMMERCIAL HANDLES

Some of the designs in this book call for purchased handles, which can come in plastic, wood, bamboo, and leather. You can always substitute commercial handles for sewn straps, if you like.

MAGNETIC CLASPS

Magnetic clasps can be found at fabric stores. They give your purse a professional-looking detail and a reliable closure, and they're easy to install—no special tools required. Just follow the manufacturer's instructions.

techniques

Refer to this section for a refresher on the various methods included in the projects in this book.

MACHINE STITCHING

If you're a novice sewer, or just haven't had your machine out in the last decade, here are some sewing basics to get you stitching.

Make sure the tension on your sewing machine is set properly for the fabric you're using. Test a scrap of fabric before working on your project. The stitches should be smooth on both sides of the fabric. If one is loose and loopy or if one is pulling too tight, consult your sewing machine manual to adjust the tension for the top thread or the bobbin. Here are the bottom-line basics for sewing a seam.

1. Pin the fabric pieces to each other with straight pins placed at right angles to the seam. Most seams are sewn with the right sides of the fabric facing each other (called "right sides together") and the raw edges aligned.

2. Hold both the bobbin and top thread securely as you make your first stitches to keep the thread from being pulled down into the machine. As you stitch, pull out the pins before they reach the needle. The pins will dull the needle if nicked, or, in the worst scenario, break it.

3. Let the machine do the work—don't pull the fabric through; just gently guide it to keep the right seam allowance as the machine pulls it through.

4. A purse can go through a lot of tugging and tossing in daily use. So, when sewing purse seams, it's helpful to reinforce your stitching at the start and end of each seam. To do so, sew the first several stitches of the seam, backstitch over them, and continue forward sewing the rest of the seam. At the end of the seam, backstitch a few stitches over the seam, and stitch forward again to the end.

To really secure the stitching, you can also tie the loose threads into a knot. Pull lightly on the bobbin thread until a loop of the top thread appears on the bobbin side of the fabric. Pull the top thread through the fabric so both loose threads are on the same side of the fabric. Tie a double knot. Repeat at the other end of the seam.

MEASURE UP

The plate under the presser foot on your sewing machine has hash marks in ⅛-inch (3 mm) increments to help you gauge the width of your seam allowance. (If they aren't labeled, measure from your needle outward to the marks to identify which one to use for your project.)

SEWING CORNERS

To sew around a corner, stitch to the corner point and stop with the needle down in the fabric. Lift the presser foot and pivot the fabric. Return the presser foot and continue sewing. This makes a nice sharp corner. You'll want to keep the sharp stitching and clip the corner fabric. Speaking of which...

CLIPPING CORNERS AND CURVES

When you sew a piece inside out, the seam allowance can bunch together when turned right side out. Luckily, getting rid of all that bulk and preventing the dreaded bunchiness is just a few snips away.

Corners—Before you turn the material right side out, clip the corner seam allowance at a 45° angle to the raw edge, near the stitching, being careful not to clip any stitches (figure 1).

Curves—For fabric on an outside curve—one that's convex—notch the seam in several places (figure 2).

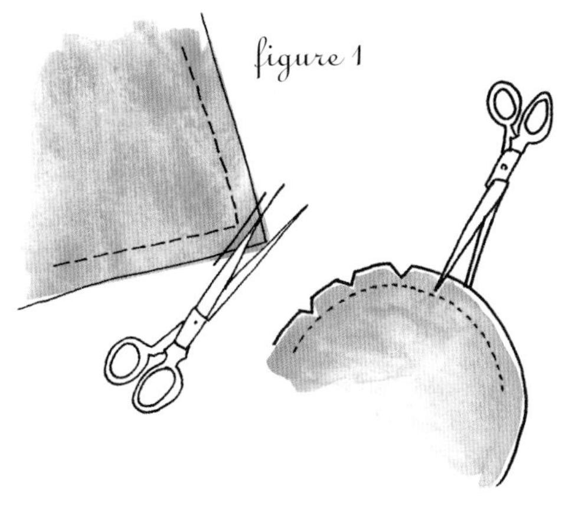

figure 1

figure 2

figure 3

The notches give the fabric room to scrunch together without a lot of overlap in the seam allowance when you turn the fabric right side out. For fabric on an inside curve—one with a concave shape—just make clips into the seam allowance (figure 3). This will allow the seam allowance to spread out a bit when you the turn the fabric right side out, giving a smoother curve.

Grading Seams—On a straight seam that joins several thicknesses of fabric, as the Bowled Over (page 92) with its patchwork inset, you may want to thin the seam allowance with graduated cuts. Trim each thickness of fabric to a different width, with the first close to the seam allowance and each successive layer a little wider.

TOPSTITCHING

Technically speaking, topstitching is a line of stitching done on the right side of the fabric, parallel to an edge or a seam. It's a decorative way to hold a panel or lining in place, to crisply flatten a seam, or purely for embellishment. Use matching thread for a subtle effect, or topstitch with a contrasting thread for visual pizzazz.

EDGESTITCHING

Much like topstitching, edgestitching is also a line of stitching on the right side of the fabric, but it is sewn closer to the fabric edge than topstitching.

BASIC PIECING

Piecing or patchwork can be symmetrical in even squares or rows, or it can be more freeform. The Patch Madame bag (page 104) uses a freestyle patchwork, applying scraps of material every which way and holding them in place by quilting.

USING A ROTARY CUTTER

Scissors and a measuring tape are all you need to cut your patchwork fabric pieces. But, when you have many pieces to cut, a rotary cutter with a self-healing cutting mat and see-through ruler can save lots of time and get you sewing sooner.

1. If the fabric has an uneven edge, straighten it by placing the ruler about ½ inch (1.3 cm) over the edge, at right angles to the straight grain of the fabric. Place the rotary cutter against the ruler on the end nearest you.

2. Holding the cutter at a 45° angle—and pressing down on the ruler with your other hand—wheel the cutter away from you along the ruler's edge.

3. With this straight edge, cut the fabric into strips the width you need. Layer four to six strips on top of each other, and cut them all to length at once. If you have to force the cutter through the layers, try using fewer pieces at a time.

SEWING PATCHWORK

When stitching symmetrical pieced patterns, such as the Bowled Over bag (page 92), use this method.

1. Arrange the patchwork pieces in the pattern you want to use. Pin the first two pieces with the right sides together and stitch the side seam. Add the next piece, with right sides together, and sew the next side seam. Continue adding pieces to the desired width of the patchwork.

2. Repeat step 1 to piece as many separate rows as desired. Arrange the patchwork rows in the order you want to sew them.

3. Pin two rows with the right sides together, and stitch the long top seam. Press the seam allowance. Add another row, with the right sides together, sew the next seam and press. Continue until all rows are joined. Pressing the seam allowances as you go ensures more accurate piecing.

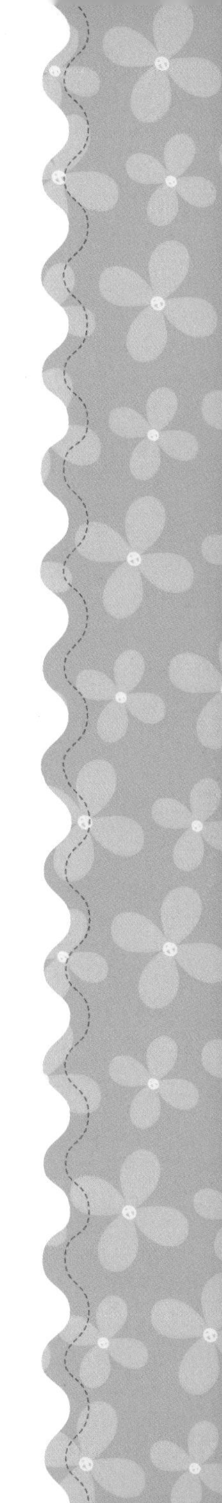

QUILTING

Quilting is the technique of stitching through layers of fabric to create a decorative, textured effect. Simply place a layer of batting (or other filler fabric) between your purse fabric and a piece of thin cotton backing fabric (quilters call this "making a sandwich"). Pin carefully, or baste with long stitches, to keep the layers in position. Then topstitch the design or pattern as desired on your machine. Use a contrasting color of thread to make your design really pop, or a matching tone for a more subtle effect. The Tea Bag pouch (page 50) uses a standard quilting pattern to achieve its look, while the Patch Madame bag (page 104) uses a more freeform style for a wackier effect.

GATHERS, RUFFLES, AND PLEATS

Pleats can give dimension and really emphasize the glory of your fabric. Simply make folds in your fabric—according to the markings on the template, if included—and stitch them down when you're sewing the seam.

Gathering, meanwhile, can be used for decoration to add a volume of fabric in a small space or can be used to create ease so you can fit a larger piece to a smaller one. A ruffle is one use of gathering—it's a strip, gathered and stitched into place. Edges can be hemmed or left unfinished to suit your style. The ruffles on the Ruffled Delight bag (page 110) are folded to provide a finished edge.

MAKING GATHERING STITCHES

1. Sew a single row of basting stitches ¼ inch (6 mm) from the fabric edge. Leave the threads free on each end of the stitching—don't backstitch or tie a knot. Sew a second row of basting stitches ¼ inch (6 mm) from the first row.

2. Grasp both top threads at one end of the gathering stitches and pull gently, cinching the fabric into gathers. Repeat from the other end of the stitching until you have the desired length or volume of gathers. Even out the gathers with your fingers.

3. To keep the gathers in place, sew a straight row of regular stitches—not basting stitches—between the rows of gathering stitches. Or, pin the gathered material to the fabric it will join and sew the seam.

4. After sewing the piece in place, remove any basted gathering stitches that show on the finished side of the bag.

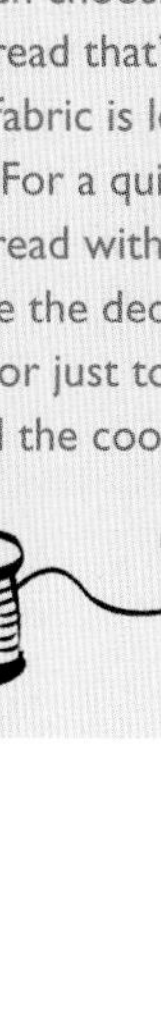

CHOOSING THREAD

When choosing thread for a project, a thread that's slightly darker than the fabric is less likely to stand out. For a quilting pattern, choose a thread with greater contrast to make the decorative element apparent, or just to announce, "Hey, look at all the cool-looking sewing I did!"

ATTACHING BIAS TAPE

Follow these instructions for sewing on bias tape (which is also called binding)—whether it's tape you made or bought from the store. If your binding goes completely around the project and meets itself at the end, such as in the Patch Madame bag (page 104), follow the instructions for Binding a Circumference. If your binding does not join, as in the Ruffled Delight bag (page 110) where the ends of the binding are hidden in a seam instead, follow the instructions for Binding a Seam.

BINDING A SEAM

1. Measure the length of the seam you will bind and add some extra for folding under the raw edges later. Cut this length of bias tape.

2. Open the folds of the tape and pin the raw edge of the tape to the raw edge of the fabric with the right sides together. Make sure to fold under both ends and pin them down. Sew the layers together, stitching in the crease of the binding (see figure 4).

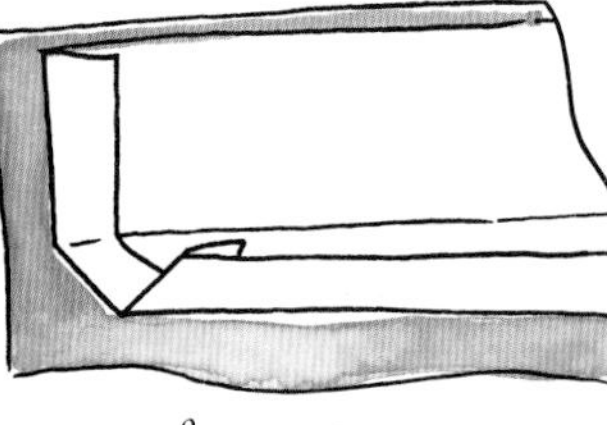

figure 4

3. Fold the bias tape over the seam allowance to the other side of the fabric. Pin in place.

4. Slipstitch by hand to secure (see figure 5), or topstitch along the edge of the tape, so the stitching hardly shows and catches the binding on the back (see figure 6).

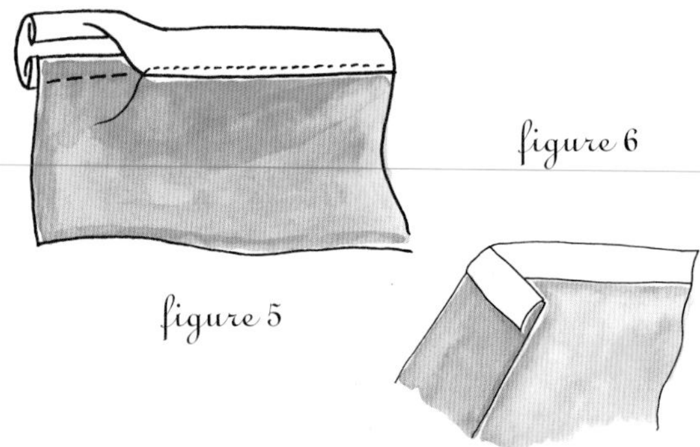

figure 6

figure 5

BINDING A CIRCUMFERENCE

Follow the instructions for Binding a Seam (at left), except after step 2, stop stitching 3 inches (7.6 cm) from the end. Clip the loose end so that 1 inch (2.5 cm) of tape overlaps the stitched tape. Continue with steps 3 and 4. Then, stop sewing 2 inches (5.1 cm) from the end (again, also the starting point) on the line of stitching you just completed. Lap the ends by folding the loose tail under ½ inch (1.3 cm), as shown in figure 7. Then finish stitching the binding down.

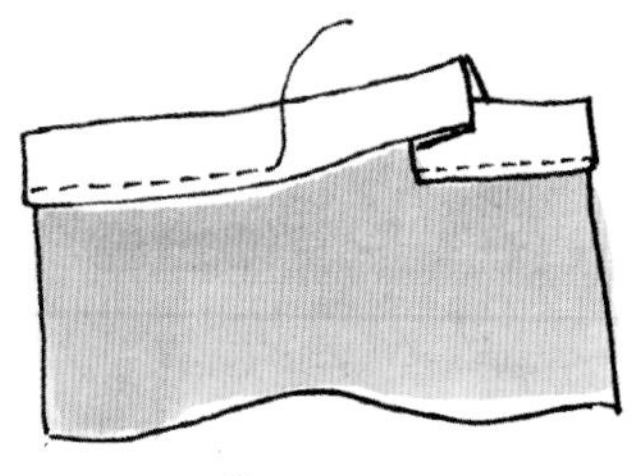

figure 7

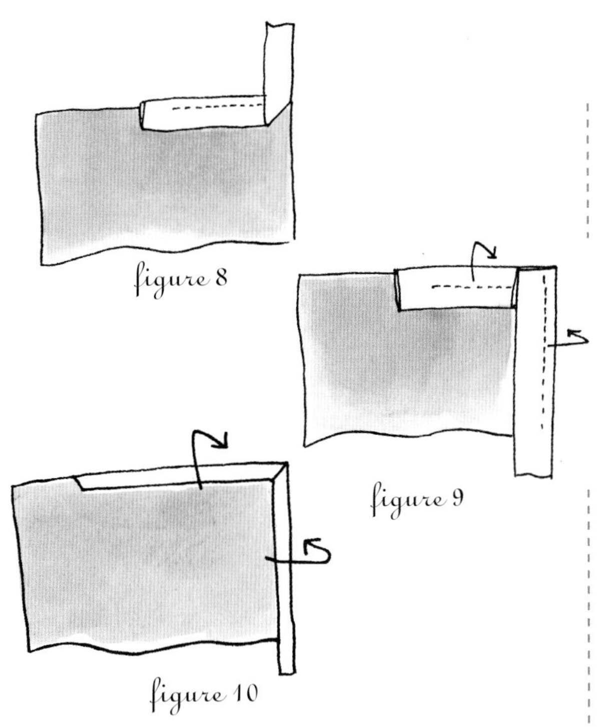

MITERING CORNERS

When binding corners, use this mitering technique for a crisp, polished appearance.

1. Bind the seam and stop ¼ inch (6 mm) from the corner. Fold the binding over itself to create a crease (figure 8).

2. Fold the binding back down. Rotate the fabric 90° and, skipping the corner, begin stitching ¼ inch (6 mm) from it toward the next corner (figure 9). Continue in this manner until you've stitched down all the binding.

3. Fold the binding over to the back side of the fabric, and fold the edges of the corners as though wrapping a package (figure 10). Hand stitch the binding to the other side.

MAKING BIAS TAPE

Purchased bias tape is convenient, but it doesn't match the delight of making the perfect finish for your bag yourself. Most of the purse projects in this book use ½-inch (1.3 cm), double-fold bias tape, so cut your initial bias strips 2 inches wide unless otherwise specified.

1. Cut strips four times as wide as your desired tape at lines running 45° to the selvage (figure 11).

2. Place one strip over the other at a right angle, with the right sides together. Stitch diagonally from one corner to the next of the overlapping squares (figure 12).

3. Cut off the corners of each seam, leaving a ¼-inch (6 mm) seam allowance. Open up the seam and press it flat. Fold the strip in half lengthwise, right side out, and press. Open the strip, and press the raw edges in to the center. You now have single-fold bias tape (figure 13).

4. To make double-fold bias tape, fold again in the center and press, as shown in figure 14.

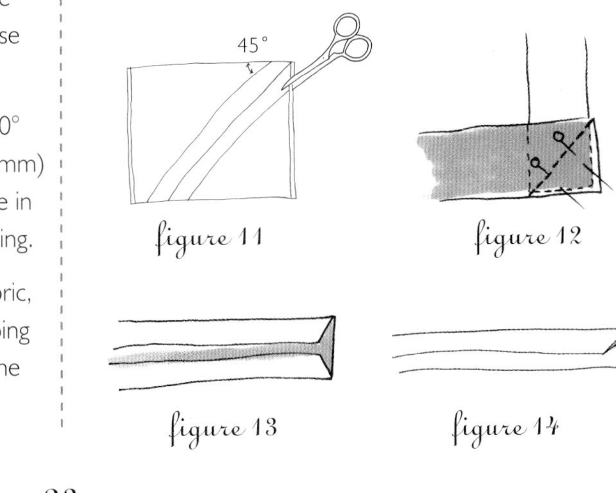

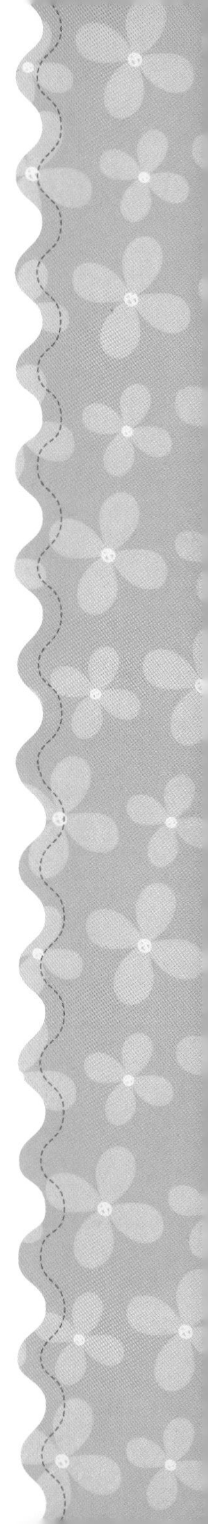

SEWING A PANEL INSERT

An insert is a section of fabric sewn in between two other pieces to add interest to the whole—such as a colorful strip sewn between two side pieces to make up a purse front, as in Just in Case (page 40). The Go Retro bag below uses a gathered insert between the top and bottom panels, and Bowled Over (page 92) uses a quilted insert between the front pocket panels.

1. Apply interfacing to the wrong side of the purse body panels if the fabric requires it.

2. With the right sides together, sew one purse body to the appropriate edge of the insert. Clip the seam, following the instructions on page 19, and press it. Repeat for the other piece.

INSERTING LININGS AND ZIPPERS

Just about every purse has a lining. It makes a bag more durable and gives a more professional finish. The purses in this book use two methods of lining. Lining A is for an open-top purse with no zipper. Lining B is for installing a zipper and a lining.

LINING A—FOR A NON-ZIPPERED BAG

1. Stitch the front and back bag lining pieces with the right sides together. Add the bottom lining piece (if there is one), stitching with the right sides together. Leave a 5-inch (12.7 cm) gap in the seam at the bottom of the bag for turning the fabric right side out later (figure 15). Trim the seams (page 19) and press them open. Leave the lining wrong side out.

2. With the right sides together, slip the lining over the bag and line up the top edges. Pin them together. If the bag has straps or a closure tab, make sure they are between the fabric and lining and not caught in the seam allowance. Match the side seams and stitch (figure 16). Trim the seam and clip the curves.

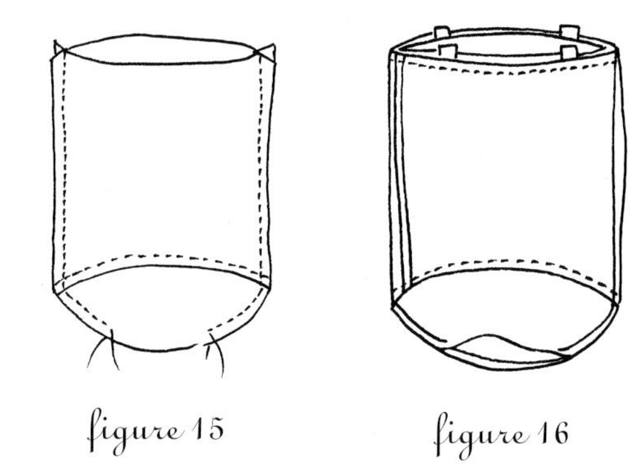

figure 15 *figure 16*

3. Turn the bag right side out through the 5-inch (12.7 cm) gap in the bottom of the lining (figure 17). Stitch the gap closed by hand or machine. Push the lining into the bag and press the top edge. Topstitch if desired.

figure 17

LINING B—FOR A ZIPPERED BAG

1. Lay the zipper wrong side up along the top edge of the purse front. Baste or pin in place (figure 18). Place the purse front lining fabric wrong side up over the zipper, aligning the top and sides with the purse front. Baste or pin in place.

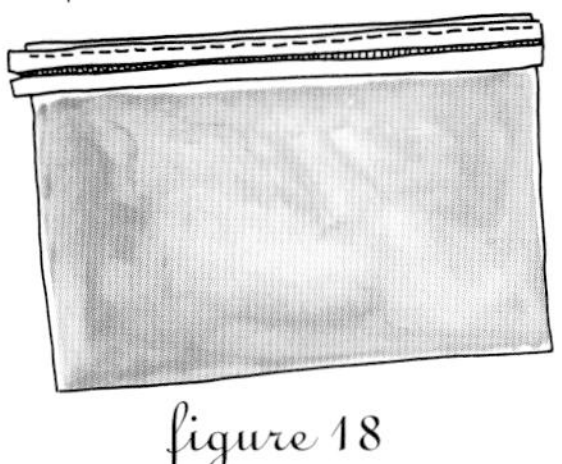

figure 18

2. Using a zipper foot, sew along the top edge, securing the zipper between the purse front and lining. Fold these pieces back over the zipper, pinning out of the way if necessary, and repeat on the other side of the zipper with the purse back and purse back lining. Flip the right sides out and topstitch along both sides of the zipper. This will keep the lining from getting caught in the zipper as you use it.

3. Open the zipper halfway. Separate the lining from the purse fabric, matching lining to lining and purse front to purse back with the right sides together. Fold the zipper in half, aligning the teeth, and make sure the zipper teeth are pointing toward the lining. Pin all of the sides in place (figure 19).

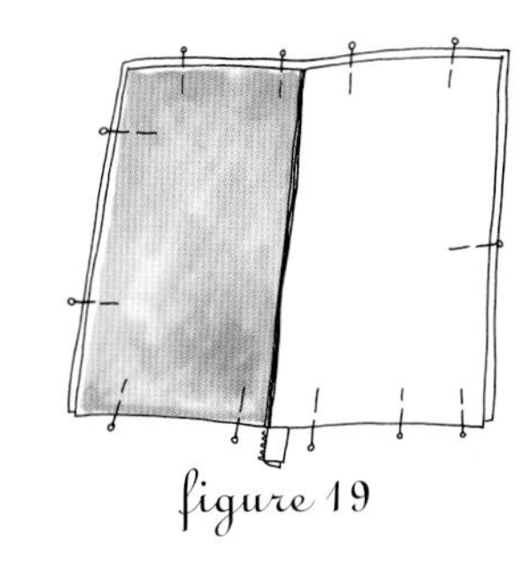

figure 19

4. Sew along all sides with a ½-inch (1.3 cm) seam allowance, leaving a gap in the bottom of the lining for turning. Clip the corners and any bulk in the seam allowance at the zipper.

5. Turn the purse right side out through the gap in the lining, pushing out the corners with a chopstick or knitting needle. Hand stitch the gap closed, and tuck the lining into the purse.

SEWING POCKETS

Whether you sew a pocket in the lining of your bag, like the Yo, Chica tote (page 99), or make it a front-and-center element of your purse, as in Miss PR (page 96) or the Checkmate bag (page 118), it's easy to do! Start with two identical cut pieces—a pocket front and pocket back.

1. Apply interfacing to the pocket pieces if the fabric requires it. Then lay the pocket front and back with right sides together and sew around them. Leave a gap on one short side for turning inside out.

2. Trim the corners according to the instructions on page 19 and turn the fabric right side out. Press the pocket flat, tucking in the raw edges of the gap even with the seam allowance.

3. Pin the pocket in place on the bag front or lining, making sure the edge with the turning gap becomes the bottom of the pocket. Stitch around the sides and bottom of the pocket to secure it.

MAKING STRAPS

If it's long, it's a strap; if it's shorter, it's a handle, but the method to make either is the same! These instructions provide two options for strapping your purse. Strap A is for a medium-weight fabric, Strap B is for very skinny straps or for stiff material that won't turn right side out easily.

STRAP A

1. To make a strap, first apply fusible interfacing to the wrong sides of the strap fabric, if it's needed. Fold the strap in half lengthwise with the right sides together. Stitch the length of the strap, pivot at the corner, and stitch across one end (figure 20). Turn the strap inside out through the open end.

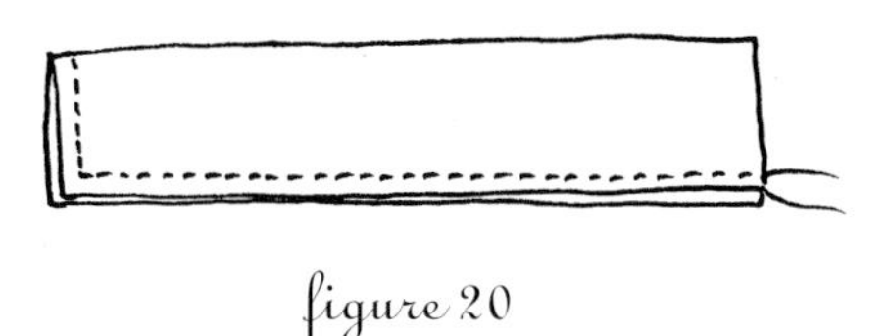

figure 20

2. Trim the seam from the short end so the strap has two open ends with raw edges. Press the strap flat with the seam in the center (figure 21).

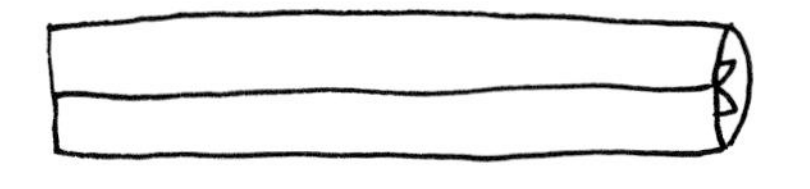

figure 21

3. For straps that will be sewn into a seam, pin the right sides of the strap to the right side of the bag (figure 22) to attach. (The seam side of the strap is the wrong side.) For other variations, such as the Dotted Bliss bag, follow the instructions for your purse (page 107).

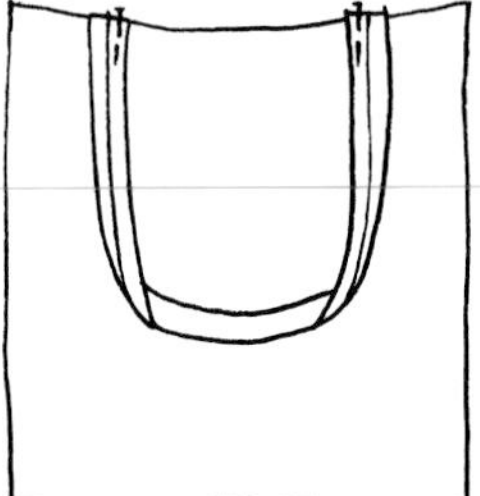

figure 22

STRAP B

1. Press the strap piece in half lengthwise, with wrong sides together, making a crease. Using the center crease as a guide, fold in each long edge to the center and press (figure 23).

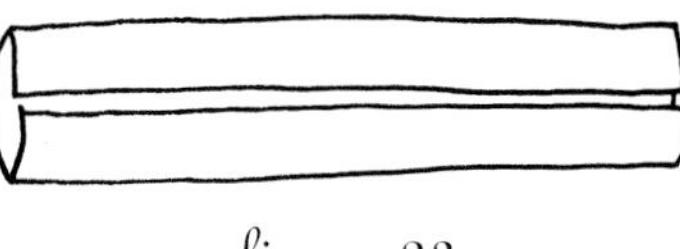

figure 23

2. Refold along the original center crease line and press again. Topstitch both long sides of the strap (figure 24).

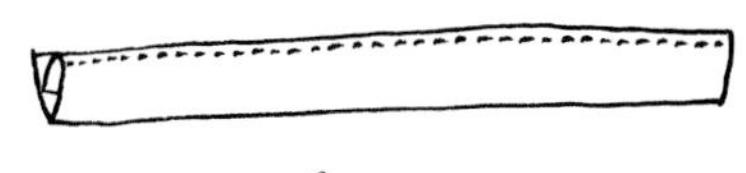

figure 24

SEWING CASINGS

A casing is a tunnel of fabric to hold a handle or drawstring. It can be the channel between two rows of stitching on the layers of the purse, like on the Happy Village pouch (page 38) and the Inside Out bag (page 115). Or another method is a casing added on with a strip of fabric, as in the Buckle Up pouch (page 74).

STRIP CASINGS

1. Fold all of the edges in ½ inch (1.3 cm) on the drawstring casing and press. Machine stitch the folds on the short edges.

2. Turn the bag inside out. Pin the drawstring casing in place. Machine stitch close to the top edge of the casing, reinforcing the stitching at the beginning and end. Repeat for the bottom edge of the casing, leaving the short ends open (figure 25).

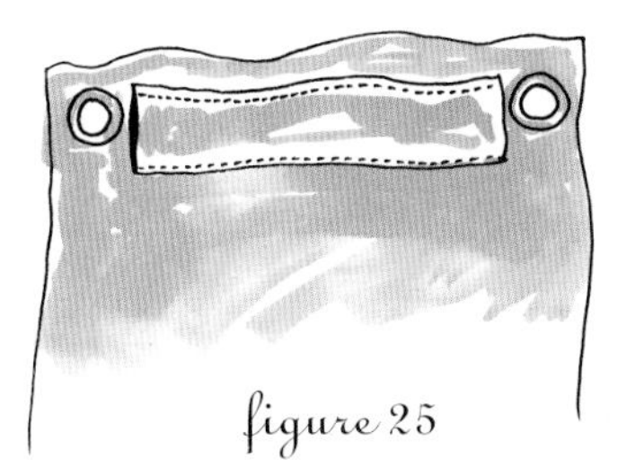

figure 25

3. Turn the bag right side out and thread the drawstrings for the bag.

THREADING A DOUBLE DRAWSTRING

Use anything from cotton cord to silky ribbon to cinch your bag closed with drawstrings. This method uses two drawstrings that are both pulled to close the bag.

1. Starting at one side seam of the bag, thread one cord into the drawstring casing (or through drawstring loops). Use a safety pin at one end of the cord to help work it through the casing, if needed. Thread the drawstring around the bag, coming back out through the same place you started.

2. Repeat with the other cord, starting from the other side seam and coming out the same place you started. Even out the ends of the cords and tie a knot on each side. For short drawstrings, slide the knot into the casing to hide it, if you like.

FELTING WOOL

Felting is the process of shrinking wool fibers to create a thick, rich texture and sturdy fabric that will not ravel. Choose garments that are 100 percent wool, and remove zippers and buttons before you begin.

1. Put your washing machine on the hottest setting, add some liquid detergent, and toss in the piece to be felted. Allow the washer to run completely through its cycle. You can add a pair of jeans or a large towel to the wash to create extra agitation and speed up the felting process.

2. When the cycle is done, check your sweater. It should have shrunk significantly. If the weave hasn't tightened enough, keep washing it until it does.

3. When the sweater is felted to your satisfaction, hang it up to dry. When it's completely dry, just cut out the seams of the sweater. Your felted wool is ready to cut to size and sew.

ROUGH IT UP

If you want a rough texture to your felted wool, throw it in the dryer to dry.

EMBELLISHING

The coolest part of sewing has got to be when you add all the charming little decorations!

SEWING APPLIQUÉ

Appliqué is a decorative fabric cut out and applied to the right side of your main fabric to add texture, contrast, and color to your purse.

1. For a fusible appliqué, apply light weight bond fusible webbing to your appliqué fabric according to the manufacturer's instructions.

2. Cut out the appliqué pattern from the fused material. Peel off the webbing backing and position the appliqué on your purse fabric.

3. Press the appliqué with an iron, following the webbing manufacturer's instructions, to fuse the appliqué to the fabric. The webbing will melt, bonding the appliqué to the purse fabric.

4. Use the appliqué stitch (page 30) to secure the edges of the appliqué by hand, or machine stitch with an appliqué stitch, or other decorative stitch that will keep the fabric edges from raveling.

APPLYING SNAPS

Snaps are an easy embellishment to add to a purse, whether you want a tight closure or just want to add some fun. From magnetic purse closures to simple snaps, find a snap kit for the style you like at a craft or fabric store. Some snaps can be put in place with just a hammer and others come with a special tool for installation. Follow the manufacturer's instructions to install your snaps. For durability, it helps to have interfacing behind the fabric that will hold a snap.

FREE MOTION EMBROIDERY

Create all sorts of decorative effects and unique designs with free motion embroidery using an embroidery hoop with your sewing machine. Check your sewing machine manual for any special instructions or settings for machine embroidery.

1. Use a thin paper on the wrong side of your fabric, and place the fabric and paper in an embroidery hoop. This will keep the fabric taut as you sew and the paper will help stabilize the fabric and keep the stitches crisp.

2. Use a darning foot on your sewing machine, if you have one. Lower the feed dogs of the machine and set your stitch to a wide zigzag.

3. Place the fabric hoop under the presser foot, and lower the presser foot (or darning foot) over your fabric. Make a few initial stitches to secure the thread and then trim off the loose ends so they don't get caught in your design.

4. Sew at a high speed, moving the hoop slowly to create the pattern you want. Keep your stitching dense to cover the fabric as you go. Be sure as you sew that your fabric stays very taut in the hoop—keep your hands on the hoop edge to prevent loosening the fabric inside the hoop.

MAKING YO-YOS

Yo-yos are little gathered rosettes that add a playful touch to your purse or bag, and they're a cinch to make.

1. Measure twice the size needed for the finished yo-yo plus ½ inch (1.3 cm) and cut out a circle of that diameter.

2. Stitch around the perimeter of the circle, ¼ inch (6 mm) from the edge, folding the fabric under as you go (figure 26).

3. Gently pull one thread to gather the edges to the center (figure 27). Stitch to secure the gathered center, and knot and trim the thread. Flatten the yo-yo with your hand.

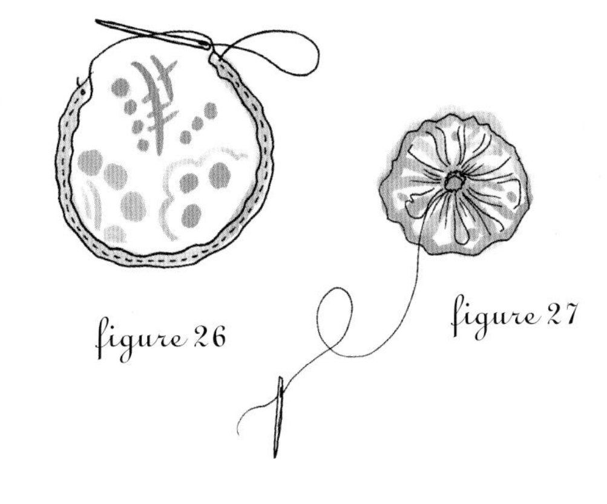

figure 26

figure 27

TIGHTEN IT UP

Shorter stitches create a larger opening in the middle of the yo-yo, while longer stitches will make a tighter center.

HAND STITCHES

Machine stitching is very likely to be your main method for sewing purses, but some may involve a few stitches by hand. Here are those you need to know for the projects in this book.

Appliqué Stitch

To camouflage the stitching for an appliqué, poke the needle through the base fabric and up through the appliqué. Bring the needle back down into the base fabric just a wee bit away. Repeat.

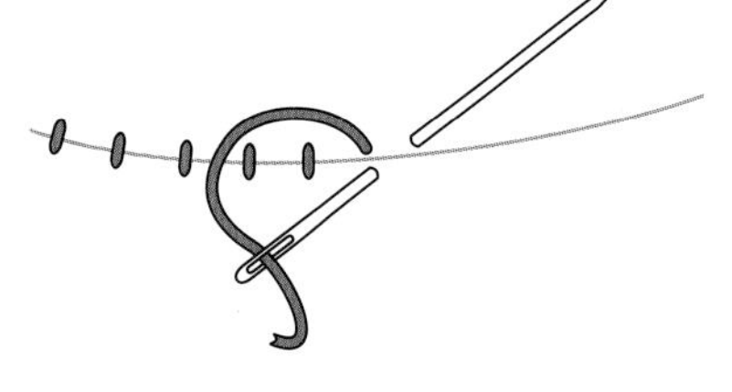

Backstitch

The backstitch is a basic hand-stitching method for creating a seam. It's good for holding seams under pressure or to outline shapes or text.

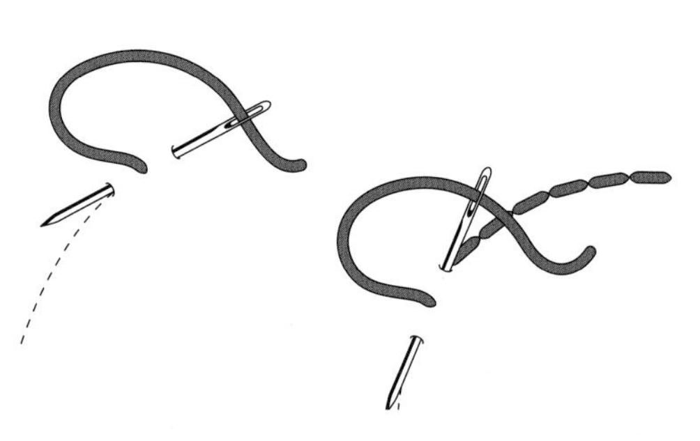

Basting Stitch

Basting is a way to temporarily secure two edges of fabric where a seam will go. The basting stitch is the same as a running stitch (see below) except you make it with very long stitches, which you can remove easily once the permanent stitch is in place. Most of the basting in the projects in this book can be done by machine, using the longest length of stitch.

Blanket Stitch

The blanket stitch is a decorative and functional technique you can use to accentuate an edge or attach a cut shape to a layer of fabric. This is the primary stitch used in the Spring Break tote (page 83).

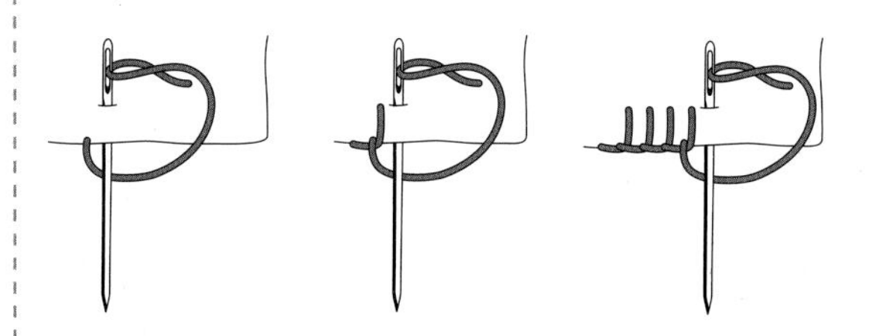

Buttonhole Stitch

The buttonhole stitch is similar to the blanket stitch but forms a knot at the fabric edge.

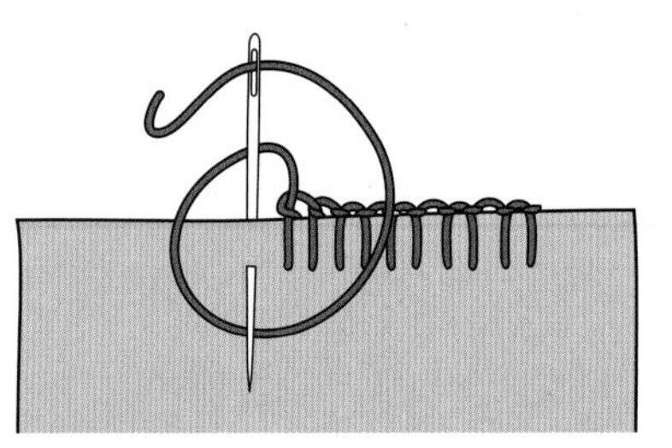

Chain Stitch

The chain stitch, or "Lazy Daisy stitch," can be worked in a circle to form a flower.

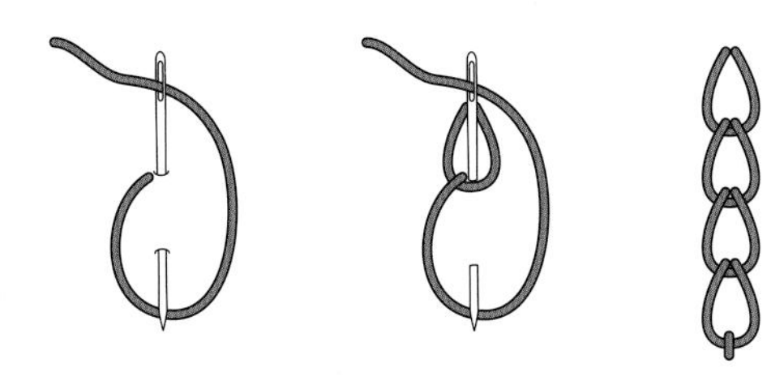

French Knot

This elegant knot, meant to be shown off, is used for embellishment. Keep the thread tight as you embroider it.

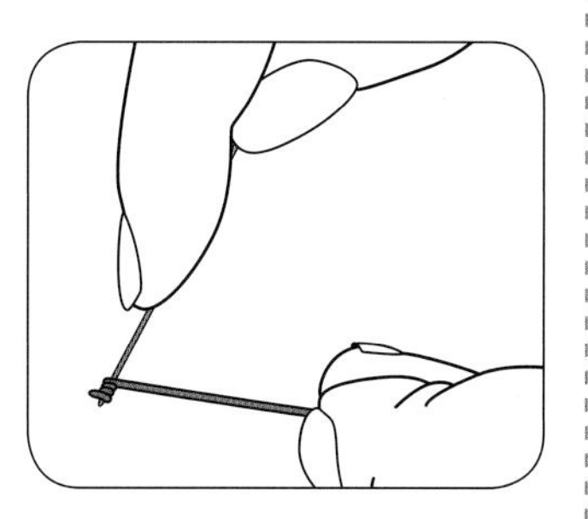

Running Stitch

You create this stitch by weaving the needle through the fabric at evenly spaced intervals.

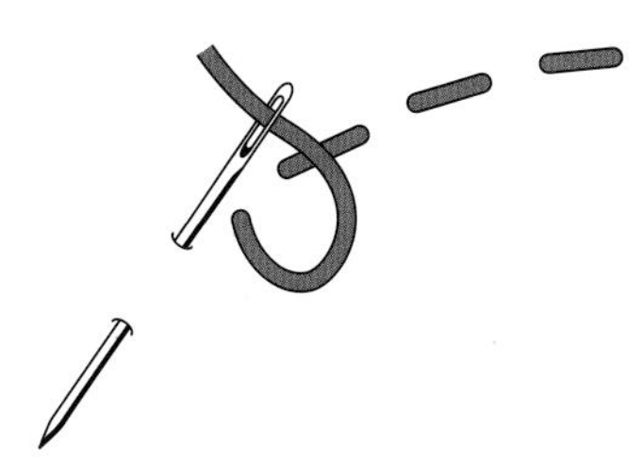

Satin stitch

The satin stitch is composed of parallel rows of straight stitches, often to fill in an outline.

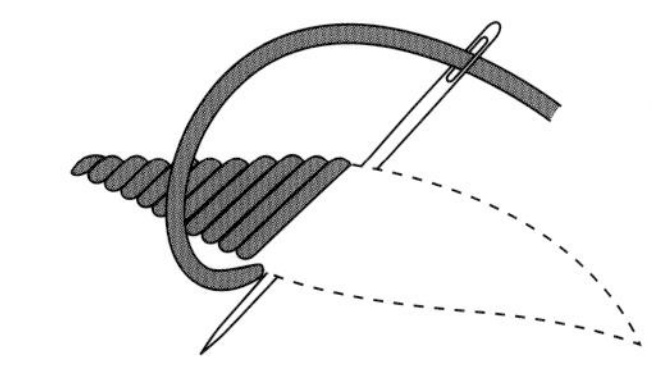

Slipstitch

The slipstitch is a good stitch for closing up seams. Slip the needle through one end of the open seam to anchor the thread, then take a small stitch through the fold and pull the needle through. In the other piece of fabric, insert the needle directly opposite the stitch you just made, and take a stitch through the fold.

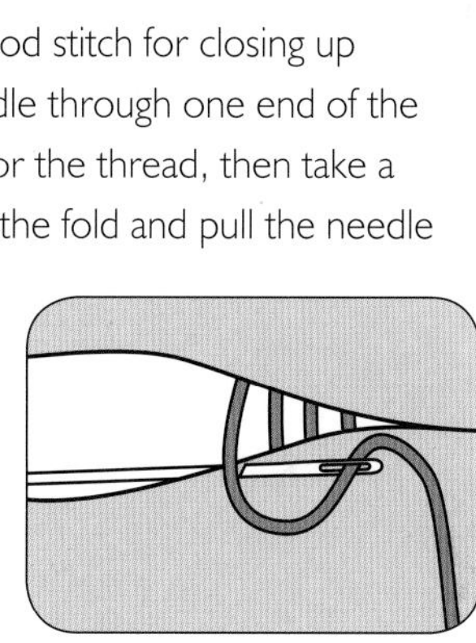

Whipstitch

The whipstitch is used to bind edges to prevent raveling. Sew the stitches over the edge of the fabric.

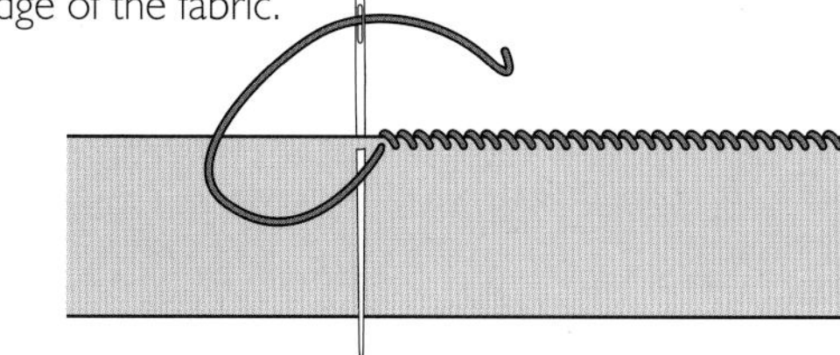

just right

Tuck the little things that matter the most in these small bags.

tokyo rose

Don't you love Tokyo Rose? With a heart as big as hers, she's simply irresistible, and she'll hold all your sweet little nothings. Slip a compact, a lipstick, or some peppermints inside, and take this little flirt out on the town.

DESIGNER

NADJA GIROD

WHAT YOU NEED

Basic Purses Tool Kit (page 11)
8 x 11¾-inch (20 x 30 cm) piece of heart print fabric
4 x 4-inch (10 x10 cm) piece of red felt
3¼ x 3¼-inch (8 x 8 cm) piece of white felt
4 x 4-inch (10 x10 cm) piece of black felt
5½ inch (14 cm) red zipper
15¾ inches (40 cm) of black cotton binding
Red, white, and black yarn

SEAM ALLOWANCE

½ inch (1.3 cm)

WHAT YOU DO

1 Enlarge and cut out the templates on page 122. Cut out the heart, face, hair, hands, and feet from appropriate felt colors. Embroider the eyes and mouth on the face.

2 Using the body template, cut out the purse front and back from the heart print fabric. Place the felt heart and face pieces on the front, and topstitch very close to the edges of each piece.

3 With the right sides together, attach the hands on both sides of the heart and the feet below the heart (figure 1). Pin the felt hair in place and attach with an appliqué stitch (page 30).

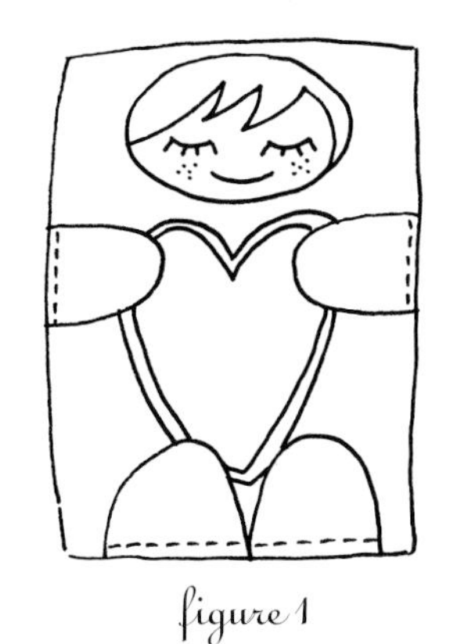

figure 1

4 Install the zipper following the instructions on page 25 for Lining B (figure 2), but omit the lining and finish the raw edges with binding instead, following the instructions on page 22 for Binding Seams with Corners.

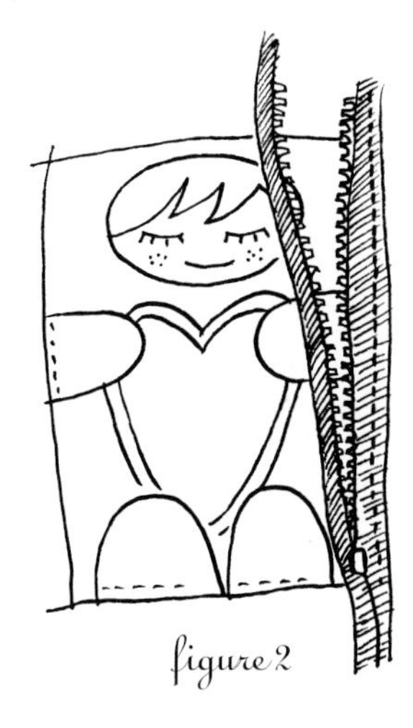

figure 2

NO RAVEL WORRIES

Felt is a great material for appliqué. You don't have to worry about binding raw edges because felt is nonwoven and doesn't ravel.

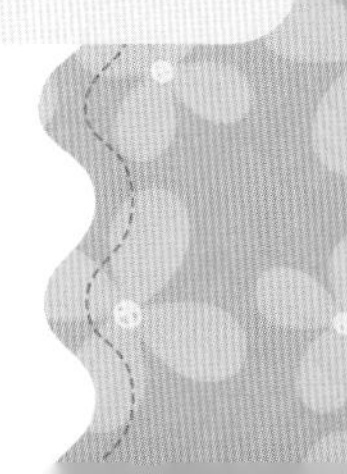

DESIGNER

AIMEE RAY

sweet pea

It's time to raid your piggy bank. Take this little coin purse to the farmer's market and you'll be singing "wee, wee, wee" all the way home. Why? Because life is fun when you use your cents.

WHAT YOU NEED

Basic Purses Tool Kit (page 11)

Pinking shears

Two 12-inch (30.5 cm) squares of felt

Two scraps of cotton fabric

Transfer paper

Embroidery needle

Embroidery floss in four shades

2½ inches (1.3 cm) of hook-and-loop tape

4 inches (10.2 cm) of peach satin ribbon

SEAM ALLOWANCE

½ inch (1.3 cm) unless otherwise noted

WHAT YOU DO

1 Using the pinking shears, cut two pieces of felt 4½ inches (11.4 cm) sqare. Cut two pieces of fabric to the same dimensions, again using the pinking shears. Using transfer paper, transfer the flower embroidery pattern from page 124 onto one piece of the felt. Embroider the design with a satin stitch, French knots, and chain stitch (pages 31).

2 Sew one portion of the hook-and-loop tape to the top of one piece of the cotton fabric. Sew the matching side of the tape to the other piece of cotton, placing it carefully to line up just right.

3 Place a fabric square and a felt square with the wrong sides together, and sew straight across the top, just above the hook and loop tape. Repeat with the other squares.

4 Line up the two sections with the right sides out and pin them together. Fold the 4-inch (10.2 cm) ribbon in half, and pin it between the layers near the top. Topstitch around the sides and bottom of the pouch.

SEW ATTENTIVE

Make your stitches carefully when sewing felt. A seam on felt fabric is harder to rip out because of the thickness and texture of the felt.

happy village

DESIGNER

AIMEE RAY

They say it takes a village, but you can create this whimsical little bag all by yourself. Be careful, though—the more you embroider, the more addicted you'll surely become!

WHAT YOU NEED

Basic Purses Tool Kit (page 11)

Transfer paper

2 circles of cotton fabric, each 14 inches (35.6 cm) in diameter

Embroidery floss in 13 shades

Embroidery needle and hoop

40 inches of satin ribbon

SEAM ALLOWANCE

½ inch (1.3 cm) unless otherwise noted

WHAT YOU DO

1 Transfer the village circle embroidery pattern from page 122 to the middle of one piece of the fabric and embroider the design.

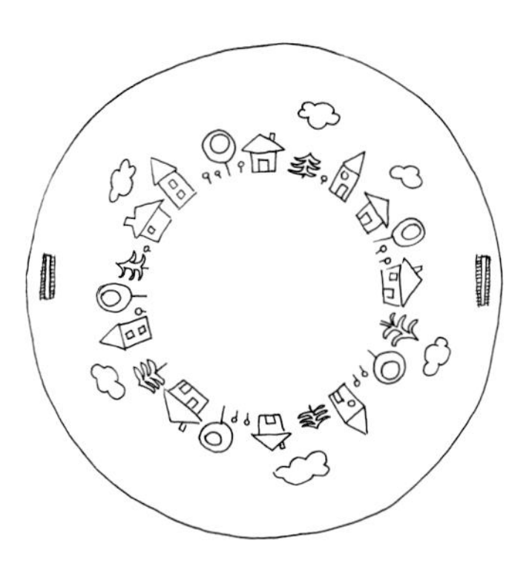

figure 1

2 Sew two buttonholes by hand with the buttonhole stitch (page 30) or by machine. Place them 1½ (3.8 cm) inches from the edge of the fabric on facing sides of the embroidered circle (figure 1).

3 Line up both circles with the right sides together and stitch around the edge, leaving a small gap for turning.

4 Turn the pouch right side out and press the seams flat. Fold the raw edges in at the gap, and topstitch around the circle, close to the edge. Sew another circle 1 inch (2.5 cm) in from the edge to make a drawstring casing, making sure to sew below the buttonholes.

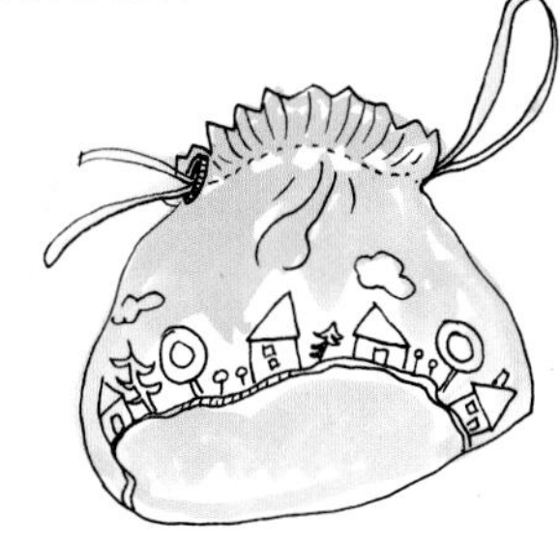

figure 2

5 Thread the ribbons as drawstrings through the casing, following the instructions on page 27 (figure 2). Sew the ends of each ribbon together and slide the ribbon around in the casing so the sewn ends are hidden inside. Pull the ribbon on both sides to close the bag.

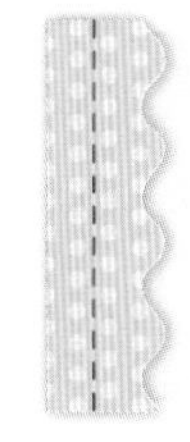

TRACE ELEMENTS

Use transfer paper to trace embroidery designs onto the fabric.

just in case

The possibilities for this charming little case are endless. Stash your glasses, makeup, office supplies, digital camera, checkbooks, change, passport, or anything else that needs a good home.

DESIGNER

AUTUM HALL

WHAT YOU NEED

Basic Purses Tool Kit (page 11)
¼ yard (22.9 cm) of linen fabric
¼ yard (22.9 cm) of cotton print for lining
6 scraps of cotton prints, 3 x 2 inches (7.6 x 5.1 cm), for patchwork
Fusible fleece
Heavyweight fusible interfacing
10-inch (25.4 cm) zipper
Thread
Fusible tape (optional)

SEAM ALLOWANCE

½ inch (1.3 cm) unless otherwise noted

WHAT YOU DO

1 Cut the fabric as described in the box at right. Arrange the 3 x 2-inch patchwork pieces in order for sewing. Join the 3-inch (7.6 cm) sides with a ¼-inch (6 mm) seam allowance. Press the seams, and trim the strip to 6 inches (15.2 cm). Sew the patchwork insert to the purse front right and left pieces, following the instructions on page 24 for Sewing a Panel Insert.

What You Cut

Linen

- *1 piece for the purse back, 6 x 10 inches (15.2 x 25.4 cm)*
- *1 piece for the purse front left, 6 x 3 inches (15.2 x 7.6 cm)*
- *1 piece for the purse front right, 6 x 5 inches (15.2 x 12.7 cm)*

Cotton print

- *2 pieces for the lining, 6 x 10 inches (15.2 x 25.4 cm)*

Fusible fleece

- *2 pieces, 6 x 10 inches (15.2 x 25.4 cm)*

Fusible interfacing

- *2 pieces, 6 x 10 inches (15.2 x 25.4 cm)*

Cotton print scraps

- *Cut the 3-inch (7.6 cm) strips into various widths, up to 2 inches (5.1 cm), for the patchwork insert*

2 Fuse the fleece to the wrong side of the purse front and purse back pieces. Fuse the interfacing to the wrong side of the cotton print lining pieces.

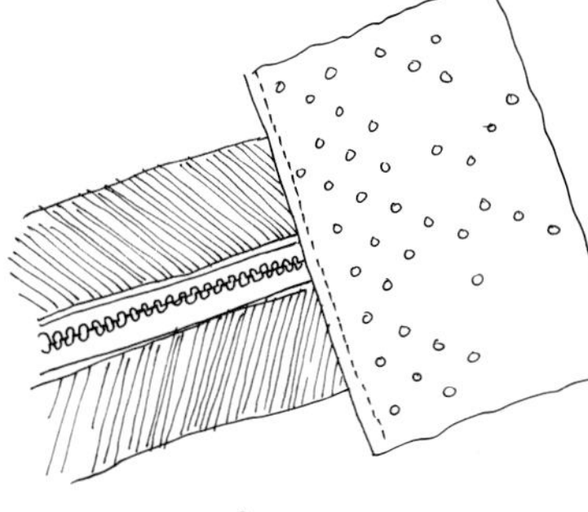

figure 1

3 Make tabs for the ends of the zipper with two scraps of fabric that are 2 inches (5.1 cm) long and the width of the zipper tape. Fold in half to 1 inch (2.5 cm) long and press. Place one tab on one end of the zipper with raw edges together (figure 1). Repeat with the other tab.

4 Install the zipper and lining following the instructions on page 25.

SLICK ZIPPER TRICK

If pins make installing a zipper awkward, fuse a piece of fusible tape to both outside edges of the zipper tape instead. Then fuse the zipper to the purse back and the purse front instead of basting. Sew the zipper in place without pins!

phone home

Your phone just called, and it wants this cozy. Snap one around a belt loop or purse strap, and you'll never misplace your cell again.

DESIGNER

AUTUM HALL

WHAT YOU NEED

Basic Purses Tool Kit (page 11)

Pinking shears

Wool felt scraps

Cotton fabric scraps for yo-yos

Thread

Embroidery floss

Snap and snap kit

Glue

Buttons

SEAM ALLOWANCE

½ inch (1.3 cm) unless otherwise noted

WHAT YOU DO

1 Use pinking shears to cut a rectangle of felt 1 inch (2.5 cm) taller and 1 inch (2.5 cm) wider than your cell phone for the front of the cozy. Clip the bottom corners into curves.

figure 1

2 Cut a piece for the back that is the same width as the piece you cut in step 1, adding a 1½ x 5-inch (3.8 x 12.7 cm) extension for the strap (figure 1). Clip the corners of the back panel and the strap into curves.

3 Cut a small leaf shape from green felt. From the cotton fabric, cut a 3-inch (7.6 cm) circle and follow the instructions on page 29 to make a yo-yo. Hand stitch the yo-yo to the front panel and sew a decorative button in the center.

4 Use green embroidery floss to make a stem for the flower. Attach the leaf by sewing a few running stitches (page 31) down the center. Make a few random satin stitches (page 31) for grass.

5 Apply the snap pieces to the top right corner of the front panel and to the end of the strap following the manufacturer's instructions. Glue a decorative button in place to cover the back of the snap on the strap.

6 Place the front and back pieces with the wrong sides together. Stitch around both sides and the bottom using a blanket stitch (page 30).

MAKE IT SNAPPY

Snap kits can be purchased at any craft or fabric store and come with all the necessary hardware for attaching them.

bling sling

DESIGNER

VALERIE SHRADER

Keep your treasures safe on the road with this well-designed jewelry pouch. The ring roll stays secure with a snap, and there are two stash pockets—one with a zipper—inside.

WHAT YOU NEED

Basic Purses Tool Kit (page 11)

3 fat quarters of different, complementary fabric (A for outside and zippered pocket; B for lining, ties, and ring roll; C for pocket and piping)

Matching thread

9-inch (22.9 cm) zipper in a complementary color

6 inches (15.2 cm) of ⅜-inch (9 mm) cording

Large hand sewing needle

Heavy-duty thread

Narrow tape

12-inch (30.5 cm) square of craft felt

1 snap fastener

SEAM ALLOWANCE

½ inch (1.3 cm) unless otherwise noted

What You Cut

Fabric A

- *1 piece for the outside, 6 x 10 inches (15.2 x 25.4 cm)*
- *1 piece for the zippered pocket, 6 x 7 inches (15.2 x 17.8 cm)*

Fabric B

- *1 piece for the lining, 6 x 10 inches (15.2 x 25.4 cm)*
- *2 strips, each 1 x 20 inches (2.5 x 50.8 cm), for the ties*
- *1 strip for the ring roll, 2 x 8 inches (5.1 x 20.3 cm)*

Fabric C

- *1 piece for the pocket, 6 x 7 inches (15.2 x 17.8 cm)*
- *2 strips, each 1½ x 10 inches (3.8 x 25.4 cm), for the piping*
- *Craft felt*
- *1 piece for the batting, 6 x 10 inches (15.2 x 25.4 cm)*

WHAT YOU DO

1 Cut the fabric as described in the box at left. Install the zipper in the 6 x 7-inch (15.2 x 17.8 cm) piece of fabric A by pressing under ¼ inch (6 mm) on each of the 6-inch (15.2 cm) edges. Center the zipper under the edges, unzip the zipper, and stitch the edges using a zipper foot (figure 1). Make the pocket by folding the fabric behind the zipper tape, forming a flat tube.

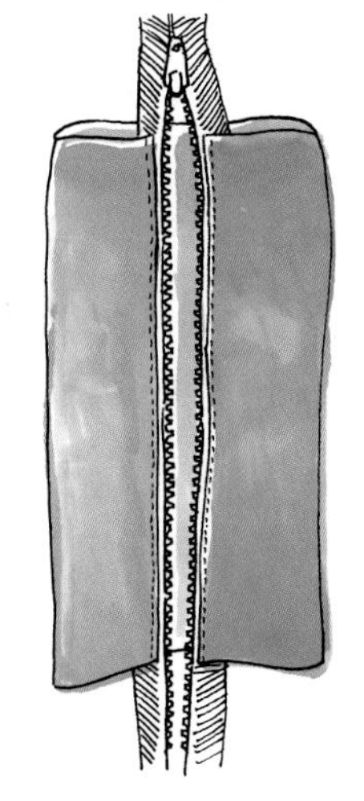

figure 1

2 Use the 6 x 7-inch (15.2 x 17.8 cm) piece of fabric B to make a pocket by folding lengthwise, right sides together, and stitch. Turn right side out and press.

3 Take the lining piece and pin the fabric A zippered pocket and the fabric B pocket in place. Stitch the bottom of the fabric B pocket in place, and stitch the top edge of the fabric A zippered pocket in place. Slide the zipper pull into the center of the pocket. Baste each pocket in place along the edges (figure 2).

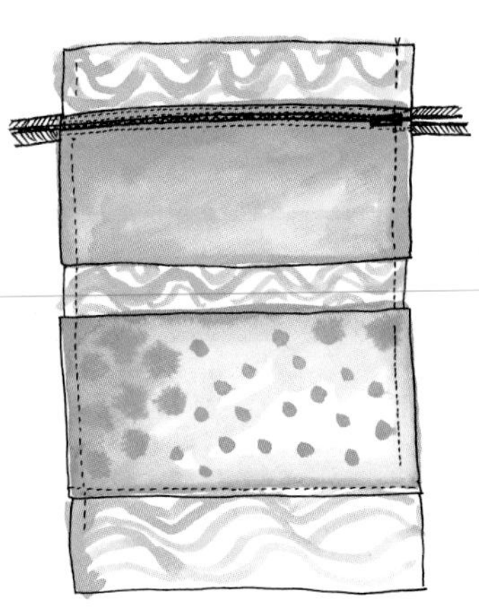
figure 2

TOTIN' TREASURE

Pieced cotton prints are pretty, but if you own jewelry worthy of an heiress, try using silks or velvets instead!

4 To make the ring roll, wrap the 2 x 8-inch (5.1 x 20.3 cm) fabric B strip around the cording, right sides together. Using the zipper foot, stitch along the cording. Slide the fabric off the cording and stitch across one end. Turn it right side out. Place a narrow piece of tape around one end of the cording. Thread the needle and tie a knot in the end of a short, doubled strand of heavy-duty thread. Stitch through the taped end of the cording. Slide the needle through the roll and out the stitched end, pulling the cording through the roll. Clip the thread and needle as close to the sewn end as possible. Trim the roll to fit inside the side seam allowances of the pouch. Trim away ½ inch (1.3 cm) from the cording inside the open end of the roll and place this end on the seam line at the right side of the lining. Sew a snap to the sewn end of the roll, placing the socket end on the lining and the ball end on the roll.

5 Make flat piping from the 1½ x 10-inch (3.8 x 25.4 cm) fabric C strips by pressing each in half lengthwise, wrong sides together. Pin one strip to each long edge of the pouch lining, placing the felt piece on the wrong side of the lining. Baste the piping in place.

6 Make the ties by folding each 1 x 20-inch fabric (2.5 x 50.8 cm) B strip in half lengthwise, pressing, and folding each edge into the crease. Press the folded edges in to the center crease and stitch, folding under each short end. Baste the ties to the center of the top edge of the lining, stacking them one atop the other.

7 Place the front and the lining together, right sides facing, and pin in place. Stitch, leaving an opening at the bottom edge. Trim the seams, clip the corners, and turn right side out. Slip-stitch the opening.

nest egg

Keep your cash stash, your plastic, and your IDs in this nifty little felt wallet. You won't want to leave your nest without it! The pinked edges show off the interior lining.

DESIGNER

AIMEE RAY

WHAT YOU NEED

Basic Purses Tool Kit (page 11)

Pinking shears

12-inch (30.5 cm) square of felt

12-inch (30.5 cm) square of cotton fabric

Transfer paper

Embroidery needle

Embroidery floss in two shades

2 inches (5.1 cm) of hook-and-loop tape

SEAM ALLOWANCE

½ inch (1.3 cm)

WHAT YOU DO

1 Using the pinking shears, cut the felt into a piece that's 4½ x 12 inches (11.4 x 30.5 cm). Do the same with the fabric. Using transfer paper, transfer the bird embroidery pattern on page 124 onto the felt. Embroider the design with a satin stitch (page 31).

2 Sew the hook-and-loop tape 2 inches (5.1 cm) from each short end of the felt on the embroidered side, measuring carefully so it matches up.

3 Place the fabric and the felt with the wrong sides together. Sew one short end together.

4 Fold the sewn end in 3 inches (7.6 cm), with the fabric on the inside. Smooth the fabric flat along the felt. Fold the other end in 3 inches (7.6 cm). Make sure the fabric is smooth inside the felt when the ends are folded in. Pin the short end together, unfold, and stitch the fabric and felt end together. Trim off the extra fabric.

5 Fold the end back again and pin both folded ends. Sew a seam on both long sides, securing the pockets. Fold the wallet in half so the hook-and-loop tape meets to close the billfold.

PINKING PRACTICE

Hold your pinking shears straight when cutting fabric, and close the blades completely to make a full cut. If a fabric is too delicate for pinking, place it on top of a scrap of sturdier fabric and pink through both layers.

tea bag

This dainty little pouch will suit you to a T. Head to your favorite cafe with it, and you'll likely end up with two for tea.

WHAT YOU NEED

- **Basic Purses Tool Kit** (page 11)
- 8 x 8-inch (20.3 x 20.3 cm) square of fabric A, cotton print for the bottom of the cup
- 5 x 8-inch (12.7 x 20.3 cm) piece of fabric B, striped cotton for the top of the cup
- 2 pieces of thin satin ribbon, each 8 inches (20.3 cm) long
- 8 x 14-inch (20.3 x 35.6 cm) piece of fabric C, pinstripe cotton for the rim and handle
- 2 pieces of polyester batting, each 8 x 8 inches (20.3 x 20.3 cm)
- 8 x 12-inch (20.3 x 30.5 cm) piece of fabric D, for the lining
- 16 inches (40.6 cm) of thin satin ribbon
- Muslin for the quilt backing and the tea tag
- Threads to match
- Chalk pencil
- 8¼-inch (21 cm) length of ⅜-inch (1 cm) thick cotton cording
- Masking tape
- Safety pin
- 7-inch (17.8 cm) zipper
- Fabric marker
- String
- Embroidery needle

SEAM ALLOWANCE

¼ inch (6 mm)

What You Cut

Fabric A

- *2 bottom pieces*

Fabric B

- *2 top pieces, with the stripes vertical*

Fabric C

- *1 strip on the bias, 2 x 8 inches (5.1 x 20.3 cm), for the handle*
- *2 strips for the teacup rim, 2 x 7½ inches (5.1 x 19 cm), with the pinstripes horizontal*

Fabric D

- *2 purse lining pieces*

Muslin

- *2 squares for the quilt backing, each 8 x 8 inches (20.3 x 20.3 cm)*
- *4 squares for the tea tag, each 3 x 3 inches (7.6 x 7.6 cm)*

DESIGNER

LAURRAINE YUYAMA

WHAT YOU DO

1 Enlarge the templates on page 127 and cut them out. Cut the fabric as described in the box at left.

2 With the right sides together, sew together one fabric A piece to one fabric B piece with a ¼-inch (6 mm) seam allowance. Press the seam flat. Position the ribbon ⅛-inch (3 mm) above the seam and stitch along the center. This is the purse back. Repeat for the purse front.

3 Using a chalk pencil, transfer the quilting lines onto the pieced sides. Layer the large muslin squares, batting, and purse back on top of each other, right side facing up. Stitch in the ditch along the seam between the top and bottom pieces to hold the layers in place. Quilt the fabric by topstitching along the chalk markings.

4 With the right sides together, line up one of the shorter fabric C strips with the top edge of the quilted side. Sew along the edge. Trim the batting, leaving ⅛ inch (0.3 cm) extra at the top edge beyond the fabric. Finger press the seam open. Repeat for the other side.

5 For the handle, use the fabric C strips cut on the bias and sew following the instructions on page 26 for Strap A, using a ½-inch (1.3 cm) seam allowance and trimming it to ⅛ inch (3 mm) before turning.

6 Wrap the end of the cording in masking tape and attach the pin to it. Run the pin through the tube, threading the cording.

7 Install the zipper and the lining following the instructions on page 25. Before sewing the side seams (step 4 of those instructions), move the lining to

WHAT YOU NEED

Basic Purses Tool Kit (page 11)

¼ yard (22.9 cm) of fabric A, a cotton print for the lining and flap

¼ yard (22.9 cm) of fabric B, a cotton print for the purse front and back

4¾ x 1½-inch (12 x 3.8 cm) scrap of fabric for the tab

Thread to match fabric B and tab

Button

SEAM ALLOWANCE

¼ inch (6 mm) unless otherwise noted

WHAT YOU DO

1 Enlarge the templates on page 124 and cut them out. Cut two purse body pieces from fabric A. Using fabric B, cut two purse body pieces for the lining and cut two flaps.

2 Make a dart at one of the bottom corners of a purse body, folding the notched shape in half on itself, right sides together and matching the short raw edges. Stitch, using a ⅛-inch (0.3 cm) seam (figure 1). Repeat at all the corners of all the purse bodies, including the linings.

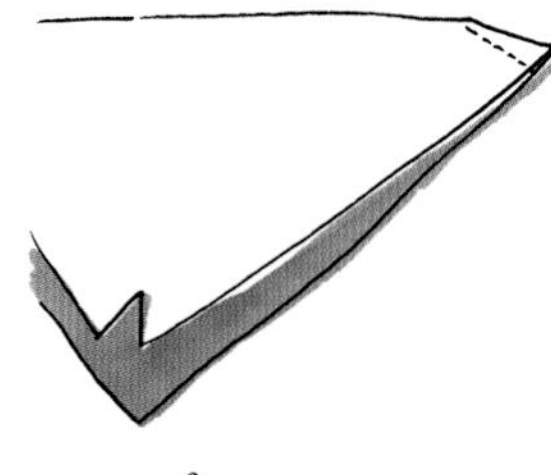

figure 1

3 Pin the fabric A purse body pieces with the right sides together and stitch the sides and the bottom. Repeat with the fabric B purse lining pieces.

4 Turn the purse right side out, and leave the lining wrong side out. Turn under the top edge of the purse ¼ inch (6 mm) and press. Turn under the top edge of the lining ¼ inch (6 mm) and press. Slide the lining into the outer pouch, aligning the seams and top edges (figure 2).

figure 2

5 Pin the fabric B flaps with the right sides together. Stitch the curved edges and leave the straight top seam open. Clip the seam, following the instructions on page 19. Turn right side out, and press flat.

6 Tuck the straight raw edge of the flap ½ inch (1.3 cm) into the back of the pouch between the lining and the purse fabric. Pin in place.

7 Make the tab by following the instructions on page 26 for Strap A. Turn the tab right side out, but leave the short end sewn. Size the hole for the button, placing it ½ inch (1.3 cm) from the tab end, and make a buttonhole on the sewn end of the tab by hand using the buttonhole stitch (page 30) or by machine.

8 Tuck the raw edge of the tab into the front of the pouch between the lining and the purse fabric, pinning it ½ inch (1.3 cm) into the pouch (figure 3). Stitch a scant ⅛-inch (3 mm) seam all around the opening of the bag where the lining meets the outer pouch.

figure 3

9 Make a ½-inch (1.3 cm) buttonhole centered on the back seam where the flap connects with the pouch (figure 4). Pull the tab through the buttonhole and over the front of the pouch. Mark the spot for the button on the front, using the tab buttonhole as a guide. Attach the button.

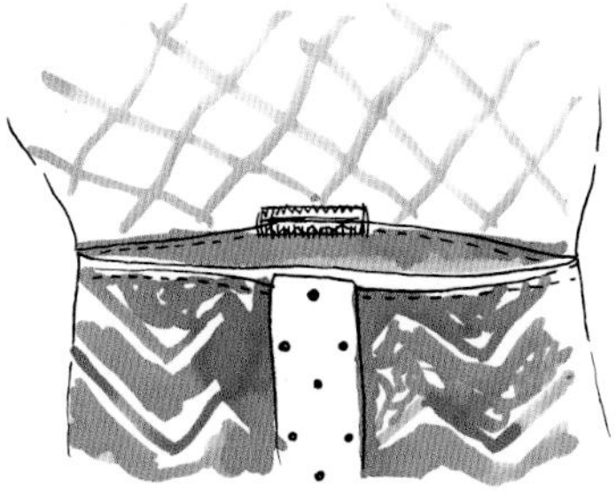

figure 4

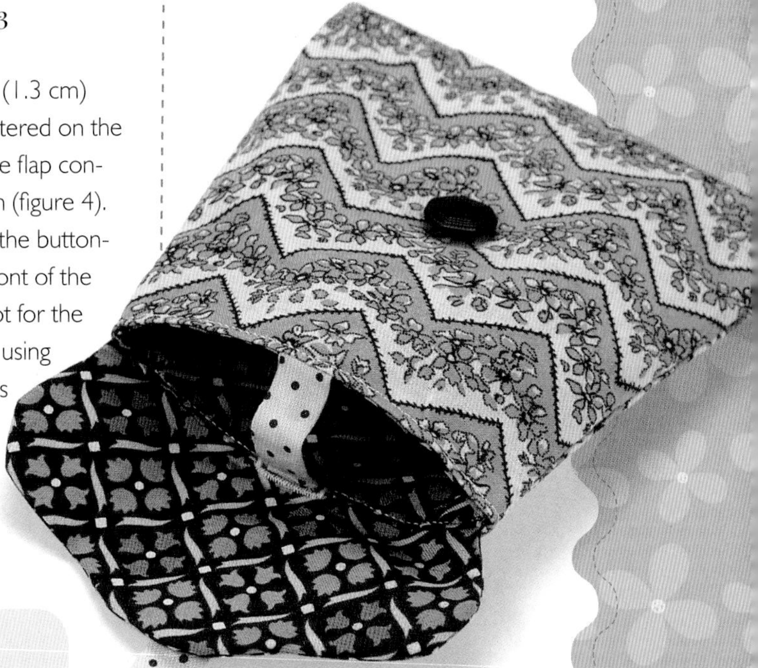

DON'T STOP NOW!

A bag this dainty begs for more embellishment. You might consider stitching on a wee bit of ribbon or adding judiciously placed lace appliqué.

bring it

Pack some extra fun in your bag
and into your day.

button clutchin'

Break out the buttons! This darling clutch snazzes up a simple print fabric with some serious button attitude. Cotton twill gives structure to the bag, and a decorative pleated pocket is the perfect spot for a compact or tickets for the show.

DESIGNER

REBEKA LAMBERT

WHAT YOU NEED

Basic Purses Tool Kit (page 11)

½ yard (45.7 cm) of twill

¼ yard (22.9 cm) of floral fabric

¼ yard (22.9 cm) of heavyweight interfacing

1 yard (91.4 cm) of ½-inch (1.3 cm) bias tape

1 small piece of hook-and-loop tape

12 buttons

SEAM ALLOWANCE

½ inch (1.3 cm)

WHAT YOU DO

1 Enlarge the templates on page 123 and cut them out. Cut the fabric as described in the box below.

What You Cut

Twill

- *2 purse body pieces*
- *2 purse lining pieces*
- *1 pocket lining*
- *1 flap*

Floral fabric

- *1 pleated pocket*
- *1 flap*

Interfacing

- *2 purse body pieces*

2 Transfer the pleat marks from the template to the pleated pocket fabric using straight pins. Working from the outside to the center, bring the outermost marks to the inner marks. Point the pleats toward the center of the clutch. Pin the pleats in place (figure 1).

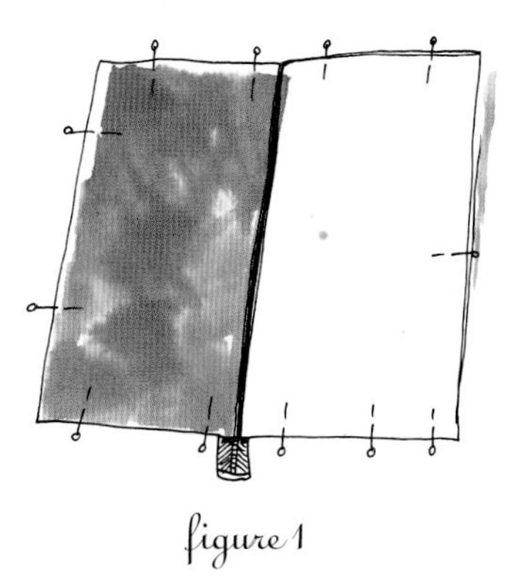

figure 1

3 Place the pleated pocket with the pocket lining, wrong sides together. Adjust the pleats if necessary to match the lining dimensions.

4 Join the pocket to the lining by encasing the top edge in bias tape, following the instructions for Binding a Seam on page 22. Hand sew the binding to the lining. Repeat with the pocket flap and flap lining, sewing along the sides and curved bottom of the flap.

5 Layer the purse pieces in the following order, from the bottom up: interfacing; purse front, right side up; pleated pocket, right side up; purse back, wrong side up; and interfacing. Line up the bottom and side edges and pin in place. Sew along the sides and bottom. Clip the corners.

6 Mark the center of the purse at the top edge of the purse back. Mark the center of the flap at the top edge. Match the purse flap to the back center with the floral side of the flap facing the right side of the purse. Pin in place along the top raw edge.

7 Use the remaining two purse body pieces for the lining, following the instructions on page 24, and topstitch ¼ inch (6 mm) from the top edge all around the purse.

8 Attach two small pieces of hook-and-loop tape to the pocket front and the underside of the flap. Sew buttons to the centers of the flowers.

FORMAL OR FUNKY

Embellished with identical understated buttons, this clutch has an elegant appearance, but you could make yours look more unconventional by using wild, mismatched buttons.

dream on

Tired of the same old purses?
Try making one from a vintage pillowcase.
With an oversize pleated bow and a
zippered closure, this clutch is
the stuff of dreams.

DESIGNER

REBEKA LAMBERT

WHAT YOU NEED

Basic Purses Tool Kit (page 11)

¼ yard (22.9 cm) heavyweight fusible interfacing

2 vintage pillowcases (or ¼ yard [22.9 m] of cotton fabric in each color)

12-inch (30.5 cm) zipper

5 inches (12.7 cm) of ribbon

SEAM ALLOWANCE

½ inch (1.3 cm) unless otherwise noted

What You Cut

One of the pillowcases (or fabric)

- *2 pieces 12 x 7 inches (30.5 x 17.8 cm)*

Other pillowcase (or fabric)

- *2 pieces 12 x 7 inches (30.5 x 17.8 cm)*
- *1 piece 14 x 12 inches (35.6 x 30.5 cm)*
- *1 piece 3½ x 4½ inches (8.9 x 11.4 cm)*

Interfacing

- *2 pieces 12 x 7 inches (30.5 x 17.8 cm)*

WHAT YOU DO

1 Cut the fabric as described in the box at bottom left. Fuse the interfacing pieces to the wrong side of the 12 x 7-inch (30.5 x 17.8 cm) pieces of fabric following the manufacturer's instructions.

2 Using the 14 x 12-inch (35.6 x 30.5 cm) rectangle, fold over 1/8 inch (3 mm) on one long edge. Fold over ⅛ inch (3 mm) again and press. Sew the narrow hem. Repeat for the other long edge.

3 Fold the 3½ x 4½-inch (8.9 x 11.4 cm) piece of fabric in half lengthwise with the right sides together. Sew the 3½-inch (8.9 cm) side together. Turn right side out and center the seam. Keeping the seam on the outside, fold the tube in half, matching the raw edges. Sew a seam along the raw edge, creating a ring. Turn the sewn ring inside out so the seam with the raw edge is on the inside of the ring.

4 Lay the large hemmed rectangle horizontally and right side up. Gather the center, accordion style, and slip the sewn ring over one end, sliding it into place in the center to create the bow for the purse front (figure 1).

5 Choose one of the rectangles as the purse front and mark its center point. Position the bow center over the center point on the purse front. Pin each corner of the bow ½ inch (1.3 cm) from the top and bottom edges of the purse front.

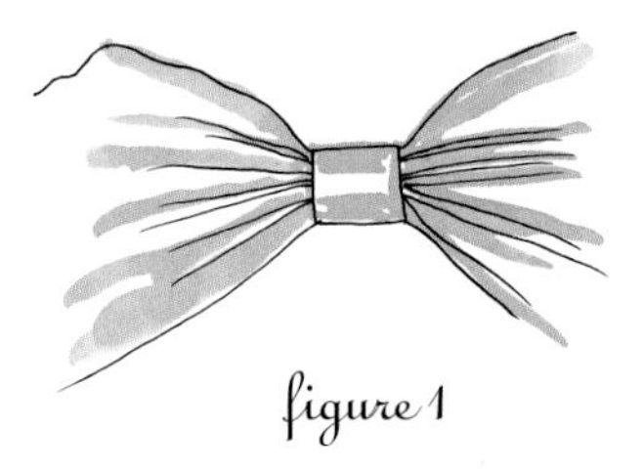

figure 1

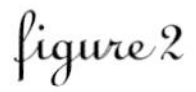

figure 2

6 Arrange the gathers of the bow along the sides of the clutch. Baste along the side edges to keep the gathers in place. Trim the excess bow fabric (figure 2).

7 Install the zipper and lining following the directions on page 25. Thread the ribbon through the zipper pull and knot it.

WHITE SALE

You can find bed linens in great retro prints for next to nothing at almost any thrift store.

pocket this

DESIGNER

WENDI GRATZ

Going to the theater or a new art exhibit downtown? This little purse is just big enough to carry all the essentials, and the straps are long enough that you can sling it across your body.

WHAT YOU NEED

Basic Purses Tool Kit (page 11)

¼ yard (22.9 cm) of crinkly cotton fabric

¼ yard (22.9 cm) of heavyweight satin

Threads to match

SEAM ALLOWANCE

½ inch (1.3 cm) unless otherwise noted

What You Cut

Cotton

- *1 rectangle, 5 x 17 inches (12.7 x 43.2 cm)*

Satin

- *1 strip for the strap, 52 x 2½ inches (130 x 6.4 cm)*
- *1 strip for the cuff, 11 x 4 inches (27.9 x 10.2 cm)*

WHAT YOU DO

1 Cut the fabric as described in the box at bottom left. Fold the cotton rectangle in half to 5 x 8½ inches (12.7 x 21.6 cm). Stitch down the long open side and across one short side for the bottom. Do not press.

2 Sew gathering stitches ½ inch (1.3 cm) in from the open end of the bag following the instructions on page 21. Divide the opening into quarters and mark each division with a dot on the right side of the fabric using chalk or a fabric-marking pen.

3 Sew the strap following the instructions on page 26 for Strap B. Topstitch on the long edges, securing all of the layers, and set the strap aside.

4 Fold the rectangle for the cuff lengthwise, with wrong sides together. Press to crease the fold, and open the fabric flat. Press under ½ inch (1.3 cm) on one long side to create a crease for sewing later, and open the fabric again.

5 Fold the cuff in half with the right sides together, matching up the short sides. Stitch the side seam of the cuff. Press the seam open, and turn right side out.

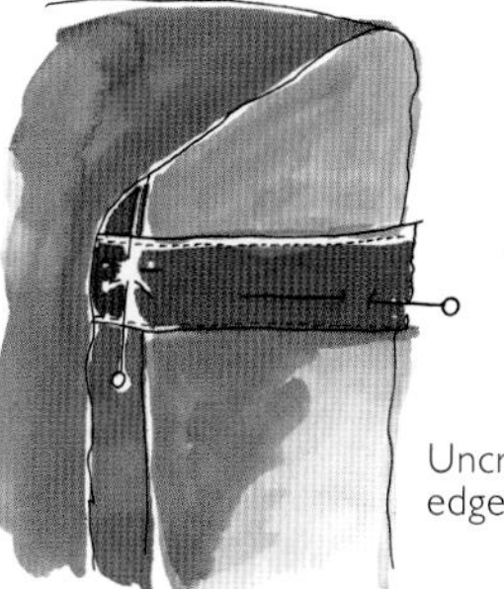

figure 1

6 Tuck the strap inside the bag, positioning it over the seam of the cuff on the right side, placing the end of the strap even with the uncreased edge of the cuff (figure 1). Sew over the

topstitching of the strap up to the center crease of the cuff. Leave the needle down in the crease, pivot, and stitch across the strap. Leave the needle down in the fabric again, pivot, and sew over the topstitching back to the raw edge. Repeat on the opposite side of the cuff.

7 Divide the cuff into quarters, and mark each division with a dot on the right side of the fabric along the uncreased edge using chalk or a fabric-marking pen.

8 Turn the cuff wrong side out. Fit the cuff over the bag, and match the dots on the cuff with the dots on the bag. Pin the cuff and bag in place with right sides together—the straps will be sandwiched between the cuff and the bag.

9 Use the gathering threads on the bag to cinch the fabric to fit smoothly to the cuff. Pin the gathers, if needed, and stitch the cuff to the bag.

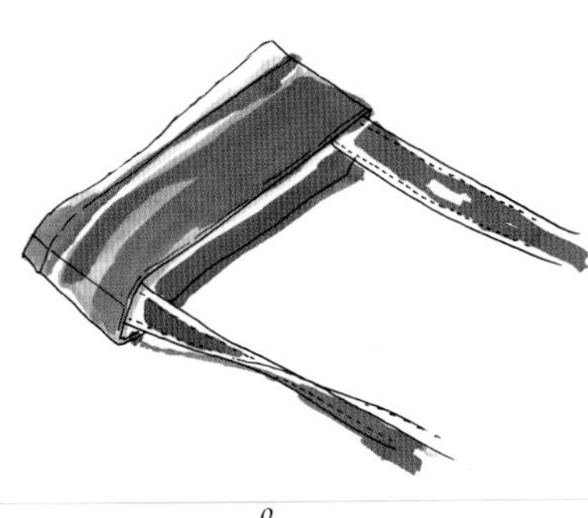

figure 2

10 Turn the cuff back up and fold along the center crease, turning the top half of the cuff to the inside of the bag (figure 2). Fold the inside edge of the cuff ½ inch (1.3 cm) under, along the crease made in step 4. Finish the inside edge of the cuff with hand stitching, covering the seam and raw edges.

KEEP THE CRINKLE

Press the crinkly fabric only where instructed, or you'll end up with flat spots on the texture of your bag.

hand bags

You have to give these bags a hand. Not only are they quirky and glamorous, but they offer a fun alternative to the plain wallet purse. Deck yours out with a sparkly button and some seed beads and let your fingers do the talking.

DESIGNER

ELIZABETH SEARLE

WHAT YOU NEED

Basic Purses Tool Kit (page 11)

½ yard (45.7 cm) of fabric A for the purse body

½ yard (45.7 cm) of fabric B for the lining

½ yard (45.7 cm) of fabric C, muslin, flannel, or batting

2 squares of fabric D for the hand appliqué, each 8 x 8inches (20.3 x 20.3 cm)

1 ½ yard (3.8 cm) of cording for the strap

8 x 8-inch (20.3 x 20.3 cm) square of heavy-duty interfacing

Button

Beads or embellishments

SEAM ALLOWANCE

½ inch (1.3 cm) unless otherwise noted

What You Cut

Fabric A

- *1 purse back and 1 purse front*

Fabric B

- *1 purse back and 1 purse front*

Fabric C

- *1 purse back and 1 purse front*

WHAT YOU DO

1 Enlarge the templates on page 127 and cut them out. Cut the fabric as described in the box at left.

2 With the right sides together, place the fabric A and fabric B fronts on top of the fabric C front and sew along the straight edge. Turn right side out with fabric C in the middle. Topstitch along the straight edge.

3 Lay the fabric B back on top of the fabric C back with wrong sides together. Place the front layers on top of the back, with right sides together. Baste all of the layers.

4 Place the cording just above the top of the front with cut ends extending 1 inch (2.5 cm) away from the seam allowance. Stitch securely. (The ends will be tucked between the lining and the purse when you sew it all together.)

5 Place the fabric A back on top of the basted back pieces and pocket with the right sides together. Sew the edges, leaving a gap for turning. Trim and clip the seams following the instructions on page 19, if necessary. Don't trim away the extra cord—it will help keep the strap from pulling out of the seam when you're using the purse.

6 Press the seams, turning under the seam allowance at the gap. Hand stitch the gap closed. Press again, turning the flap down.

7 Layer the squares of fabric D with the interfacing in the middle and the top fabric with the right side facing up. Pin the layers together. Trace a hand onto the fabric. With matching thread, stitch along the tracing line.

8 Satin stitch by hand or by machine along the hand outline (figure 1). Cut around the stitching, being careful not to cut through the stitching. Attach the hand to the bag flap with a straight stitch, sewing through the flap. Embellish it with a button for a ring and strings of beads for bracelets.

figure 1

LEFT, RIGHT, LEFT, RIGHT

Trace your own hand for this project, or draw a hand shape, and use a photocopier to reduce the drawing to the right size for the purse.

eco chic

This little tote is the most! Not only is it made using a recycled felted sweater, but it has a double-pull drawstring mechanism using black leather lacing.

WHAT YOU NEED

Basic Purses Tool Kit (page 11)

¼ yard (22.9 cm) of fabric A, cotton print for the top row of the pouch and the lining

⅛ yard (11.4 cm) of fabric B, cotton print for the middle row of the pouch

15 x 9-inch (38.1 x 22.9 cm) piece of fabric C, felted wool for the bottom row and base of the pouch

18 inches of cotton twill tape

Coordinating embroidery floss

Embroidery needle

Coordinating thread

62 inches (155 cm) of thin leather lacing

SEAM ALLOWANCE

¼ inch (6 mm) unless otherwise noted

DESIGNER

CASSI GRIFFIN

WHAT YOU DO

1 Enlarge the templates on page 128 and cut them out. Cut the fabric as described in the box below.

What You Cut

Fabric A

- *1 body-1*
- *2 purse linings*
- *1 purse base for the lining*

Fabric B

- *1 purse body-2*

Fabric C

- *1 purse body-3*
- *1 purse base*

Interfacing

- *2 pieces 12 x 7 inches (30.5 x 17.8 cm)*

2 Working with purse body-1, -2, and -3 pieces, stitch fabric A to fabric B along one long edge with the right sides together. Press the seam open. With the right sides together, stitch the other long edge of fabric B to fabric C. Press the seam toward fabric B.

3 Make a tube by folding the sewn piece in half lengthwise with the right sides together, matching the seams. Pin in place and stitch. Press the seam open.

4 Turn over the top edge of the bag—fabric A—¼ inch (6 mm) and press. Turn the piece right side out.

5 Cut the twill tape into six 3-inch (7.6 cm) pieces. Fold each 3-inch (7.6 cm) piece of cotton twill tape in half to 1½ inches (3.8 cm). Pin the loops to the right side of the bag material with raw edges even, starting at the pouch seam and spacing the loops evenly.

6 Follow the instructions on page 24 for Lining A to insert the lining, making sure the drawstring loops are sewn between the bag fabric and the lining. Using six strands of embroidery floss and an embroidery needle, blanket stitch the felted wool base to the bottom of the pouch.

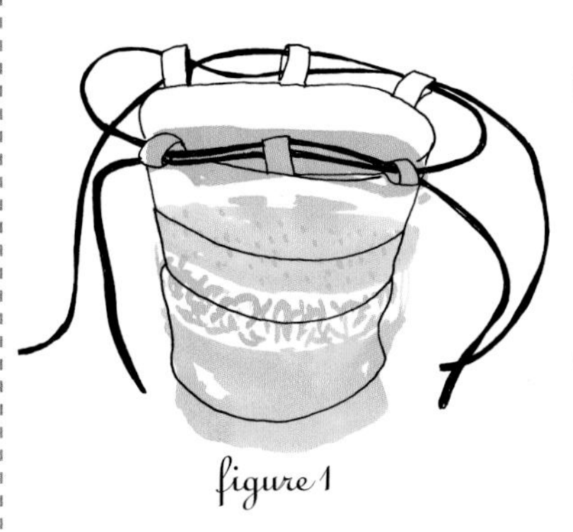

figure 1

7 Cut the lacing in half to yield two 31-inch (77.5 cm) laces. Thread the drawstring laces following the instructions on page 27 (figure 1).

FIY (FELT IT YOURSELF)

Create your own felted wool for the base of this bag from an old sweater, following the instructions on page 28 for Felting Wool.

go retro

This glamorous little bag is in a class all by itself. The handle is made with ceramic beads so pretty you'll want to slip the purse on your wrist as if it were a bangle.

DESIGNER

JOAN K. MORRIS

WHAT YOU NEED

Basic Purses Tool Kit (page 11)

¼ yard (22.9 cm) of fabric A, a microfiber

¼ yard (22.9 cm) of fabric B, a cotton floral print

¼ yard (22.9 cm) of fabric C, for the lining

1 screw-off metal purse handle

8 or more ceramic beads with holes large enough to fit onto the purse handle

1 wooden bead with a large hole

SEAM ALLOWANCE

½ inch (1.3 cm) unless otherwise noted

What You Cut

Fabric A

- *four 5 x 7-inch (12.7 x 17.8 cm) pieces for the purse body, and two 2 x 4-inch (5.1 x 10.2 cm) pieces for the handle tabs*

Fabric B

- *two 5 x 14-inch (12.7 x 35.6 cm) pieces for the insert*

Fabric C

- *two 7 x 12-inch (17.8 x 30.5 cm) pieces*

WHAT YOU DO

1 Cut the fabric as described in the box at bottom left. Sew gathering stitches on the long edges of the fabric B insert pieces, following the instructions on page 21. Gather each piece to 7 inches (17.8 cm).

2 Attach the sewn inserts to the fabric A purse top and bottom pieces for the front and back, following the instructions on page 24 for Sewing a Panel Insert.

3 Place the sewn purse front and purse back with the right sides together and pin along the side and bottom edges, making sure to line up the seams on the side. Machine stitch around the sides and bottom. Clip the corners (page 19) and turn the purse right side out. Press the sides and bottom, using a low setting with steam to protect the microfiber. Press lightly.

4 Fold in the long edges of the fabric A handle tabs, sizing them to fit through the bottom of the purse handle. Press and stitch each strip. With the right sides together, pin the handle tabs to the side seams of the bag. Match the raw edges and baste in place (figure 1).

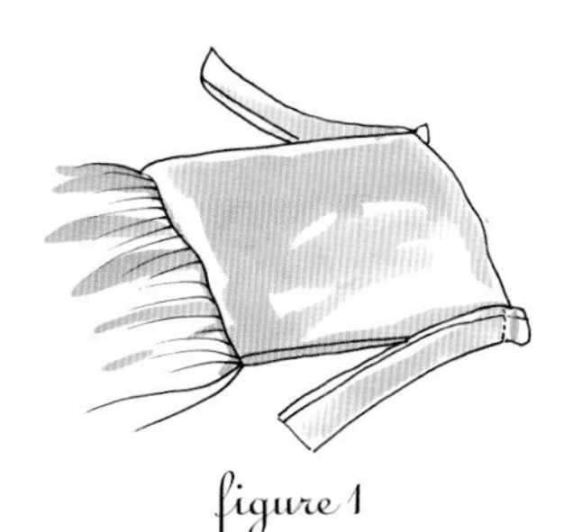

figure 1

5 Attach the lining to the bag following the instructions on page 24, with the handle tabs sandwiched between the purse fabric and the lining. Press the top edge flat.

6 Unscrew the handle base and thread the beads onto the handle. Replace the base. Run one handle tab through the handle base to the inside of the bag, and hand stitch the tab down, folding under the raw edge. Repeat with the other handle tab. Take care to stitch them well so the handle will be secure.

SHEER DELIGHT

Dress up this bag another notch by using a super diaphanous material for the midsection. It's sure to attract attention, yet the lining will keep the bag contents hidden from view.

buckle up

You hold the purse strings with this elegant microfiber bag. Featuring a spunky purse buckle and velvet ribbon, it's a bag you'll be drawn to immediately.

DESIGNER

JOAN K. MORRIS

WHAT YOU NEED

Basic Purses Tool Kit (page 11)

¼ yard (22.9 cm) of fabric A, a microfiber

¼ yard (22.9 cm) of fabric B, a cotton floral

2 silver eyelets, ¼-inch (6 mm), and setter tool

2½ yards (6.4 cm) of cording

Large safety pin

3-inch (7.6 cm) purse buckle

18 inches (45.7 cm) of 1¼-inch (3.2 cm) velvet ribbon

SEAM ALLOWANCE

½ inch (1.3 cm) unless otherwise noted

What You Cut

Fabric A

- *2 pieces for the purse bottom, each 8 x 4¾ inches (20.3 x 12 cm)*

Fabric B

- *2 pieces for the purse top, each 8 x 7¼ inches (20.3 x 18.4 cm)*
- *2 pieces for the lining, each 8 x 12 inches (20.3 x 30.5 cm)*
- *2 pieces for the drawstring casing, each 2 x 8 inches (5.1 x 20.3 cm)*

WHAT YOU DO

1 Cut the fabric as described in the box at bottom left. With right sides together, pin the fabric A purse bottom to the fabric B purse top along the 8-inch (20.3 cm) side. Stitch the seam. Repeat with the other purse bottom and purse top pieces. Press the seams open. These are the purse front and the purse back.

2 Place the purse front and purse back with the right sides together, matching the seams. Pin the bottom and sides of the purse and stitch. Turn right side out.

3 Sew the lining to the bag following the instructions on page 24. Following the manufacturer's instructions, place an eyelet 2 inches (5.1 cm) from the top on each seam of the bag, punching through all of the layers.

4 Sew two casings for the drawstrings, following the instructions for Strip Casing on page 27. Machine stitch the casings between the eyelets on each side of the bag, leaving the eyelets uncovered (figure 1).

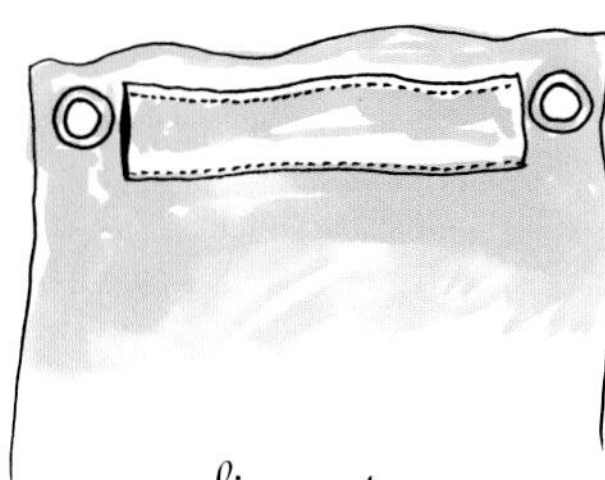

figure 1

5 Cut the cord in half and thread the drawstrings following the instructions on page 27.

6 Slide the buckle onto the ribbon and to its center point. Place the buckle on the purse front, over the seam. Pin the ribbon in place on the front and back of the purse, folding one end of the ribbon over the other at the back. Machine stitch on both long edges of the ribbon. Hand stitch the ribbon above and below the buckle.

heavy metals

With a smart geometric construction, and sparkling stitching and sequins, this look is solid gold. Take it with you to the opera and you'll hear everyone singing its praises.

DESIGNER

VALERIE SHRADER

WHAT YOU NEED

Basic Purses Tool Kit (page 11)

1 yard (91.4 cm) of silk dupioni

¼ yard (22.9 cm) of lining fabric

¼ yard (22.9 cm) of fusible batting

2 yards (1.8 m) of ¼-inch (6 mm) cording

Metallic thread in 3 different colors

Beads, sequins, etc.

SEAM ALLOWANCE

½ inch (1.3 cm) unless otherwise noted

What You Cut

Dupioni

- *one 8-inch equilateral triangle for the bottom, and three 7-inch squares for the sides*

Lining fabric

- *one 8-inch equilateral triangle for the bottom, and three 7-inch squares for the sides*

Batting

- *one 8-inch equilateral triangle for the bottom, and three 7-inch squares for the sides, but trim away the ½-inch seam allowances on each piece*

WHAT YOU DO

1 Cut the fabric as described in the box at bottom left. Trim off ½ inch (1.3 cm) from each corner of each equilateral triangle. At the center point of the trimmed edges, mark ½ inch (1.3 cm) into the seam allowance.

2 Apply the fusible to the wrong side of the pieces of dupioni.

3 Embellish the outside of the bag as desired, using metallic thread, beads, and sequins. This bag features lines of beading that cascade down the sides from a central point. Be careful to keep the beads or sequins out of the seam allowance.

4 With the right sides facing, stitch the sides to the triangular bottom, starting and stopping stitching at the marked dots (figure 1). Now stitch the sides to one another, with the right sides facing (figure 2).

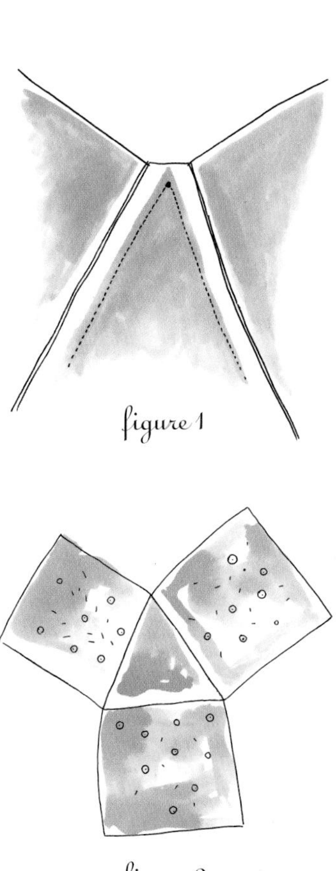

figure 1

figure 2

5 To make the straps, cut enough 1½-inch (3.8 cm) bias strips to cover three 24-inch (61 cm) straps. You can use your favorite method of covering the cording with the bias strips; in this project, the bias strips were sewn around the cording with the right sides together and the bias strip was then pulled back over the cording. **Note:** This method requires twice as much cording as called for in the materials list, as you begin in the center of a long piece of cording, stitch the bias strips across one end and then down the length of the strip using your zipper foot before you slide the fabric back over the cording.

6 Trim all the straps to the same length, if necessary. Slide the fabric back on one end, trim away about 1 inch (2.5 cm) of cording, and slide the fabric back to create an "empty" end. Pin this end to the outside of the bag at the center point of each side, placing the end of the cording at the seam line and the raw edges together. Make sure no cording extends into the seam allowance. Baste.

7 Construct the lining as you did the bag in step 4, but don't add any fusible batting.

8 Turn the lining inside out and place the bag inside the lining, with the right sides together. Make sure that the straps are between the bag and the lining. Pin and stitch around the top, leaving an opening to turn.

9 Turn the bag inside out and slipstitch the opening. Use one of the metallic threads to add a row of running stitches around the top to secure the lining.

10 To make the slider, cut a 4 x 4-inch (10.2 x 10.2 cm) piece of dupioni. Fold it in half and stitch in a ½-inch (1.3 cm) seam around one short end, and a ¼-inch (6 mm) seam around the long side, leaving the other short end open. Turn right side out through the open end and press. Topstitch along the long edges.

11 Fold the slider in half lengthwise and stitch the edges together as desired to form a flat tube. (If you want to embellish the slide, it's easiest to add decorations before you stitch the slider together.) Push the three straps through the slider.

12 To complete, finish the raw ends of the straps by sliding each strap back and trimming away about ⅜ inch (1 cm) of the cording. Fold the fabric edges under and stitch together using tiny slipstitches. Tie the straps into a knot.

As a variation, you could use a big bead instead of the fabric slider.

SILK SURPLUS

You'll have some dupioni left for another project.

snap attack

This little purse proves that two snaps are better than one, and that two pockets are where it's at. The designer used Japanese silk from an old kimono, showing that your next burst of inspiration might be as close as your closet.

DESIGNER

MICHA MAE MELANCON

WHAT YOU NEED

Basic Purses Tool Kit (page 11)

⅓ yard (30.5 cm) of fabric A, a Japanese silk

12 x 12-inch (30.5 x 30.5 cm) scrap of fabric B, a narrow-wale corduroy

⅓ yard (30.5 cm) of fabric C, a print cotton for lining

½ yard (45.7 cm) of lightweight fusible interfacing

⅓ yard (30.5 cm) of light- or medium-weight, non-fusible interfacing

20 inches (50.8 cm) of ½-inch (1.3 cm), double-fold bias tape

2 mother-of-pearl snaps

Snap application tools

SEAM ALLOWANCE

¼ inch (6 mm) unless otherwise noted

WHAT YOU DO

PIECING

1 Enlarge the template on page 124 and make three copies. Cut two out as they are, but turn one over so one template has a left flap and one has a right flap. For the third template, cut off the flap—this is the purse front.

2 Cut fabric A into three pieces at least 1 inch (2.5 cm) larger all around than the template; cut the same of fusible interfacing. Cut three identical pieces of fusible interfacing. Cut an additional piece of interfacing using the fabric B piece as a pattern. Apply the interfacing pieces to the wrong side of each fabric piece.

3 Working with a fabric A piece for the purse front, cut the piece in two at an angle. Cut a tapered strip of fabric B at least the length of the diagonal cut of silk. Cut the length against the wale so the ribbing runs horizontally across the width.

4 Position the tapered strip along one diagonal cut of fabric A with the right sides together. Stitch along the edge. On the wrong side of the fabric, press the seam allowance toward fabric A. Topstitch close to the seam on the right side of fabric A. Press again if needed. Attach the strip to the other piece of fabric A in the same way.

5 Repeat step 4 with the fabric A piece for the left flap panel and right flap panel. Make a second cut in those pieces as well, inserting a fabric B strip across the area for the flap itself.

ASSEMBLING

1 Cut one of each template (purse front, left flap panel, right flap panel) from the pieced materials. Cut one of each template from the fabric C lining, and cut one of each template from the non-fusible interfacing.

2 Place the fabric A panel right side up, position the fabric C lining wrong side up over it, and place the front panel interfacing on top of that. Sew across the top edge. Turn right side out and press the seam. Topstitch along the edge, close to the seam.

3 Repeat step 2 for the right flap panel pieces and the left flap panel pieces. Sew along the top edge, following the curve of the flap. Notch the curved edges and corners of each flap following the instructions on page 19 for Clipping Seams and Corners. Turn right side out and press. Topstitch along the top edge.

4 After completing all three pieces, position the left flap panel with the silk side up. Place the right flap panel on top of it with the silk side up. Place the front on top of that with the fabric C lining side up. Carefully line up all straight edges. Sew the three panels together along the unfinished edges of the bottom curve. Trim any irregular edges.

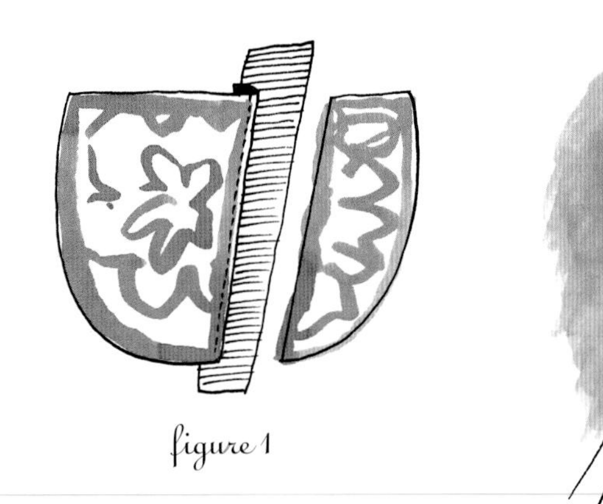

figure 1

figure 2

5 Cut a 20-inch (50.8 cm) strip of ½-inch (1.3 cm) double-fold bias tape (figure 1). Bind the unfinished curved seam allowance just sewn and miter the top corners following the instructions on page 23 for Mitering Corners (figure 2).

6 Apply a snap to each flap following the manufacturer's instructions. Since the flaps overlap, start with the right flap, which is underneath. Apply the top snap to the flap. Fold the flap over and trace the tip of the stud with a fabric pencil. Use this mark to apply the other half of the snap. Repeat for the left flap.

TOUCHY FEELY

This purse uses silk and fine-wale corduroy for a great textural contrast. Use other fabrics if you like, but to add interest make sure they have different textures or sheen.

spring break

This cheerful little clutch will put a bounce in your step. You don't even need a sewing machine for this one—the wool felt purse and flowers can all be sewn by hand.

DESIGNER

CASSI GRIFFIN

WHAT YOU NEED

- **Basic Purses Tool Kit** (page 11)
- 12 x 15 inches (30.5 x 38.1 cm) each of wool felt in gray and turquoise
- 5 x 5 inches (12.7 x 12.7 cm) each of wool felt in yellow, orange, and red-orange
- 4 x 4 inches (10.2 x 10.2 cm) each of wool felt in green
- 3 x 3 inches (7.6 x 7.6 cm) each of wool felt in lime and brown
- Embroidery flosses to match
- Embroidery needle
- 11 x 7-inch (27.9 x 17.8 cm) piece of plastic canvas
- Scalloping scissors (optional)

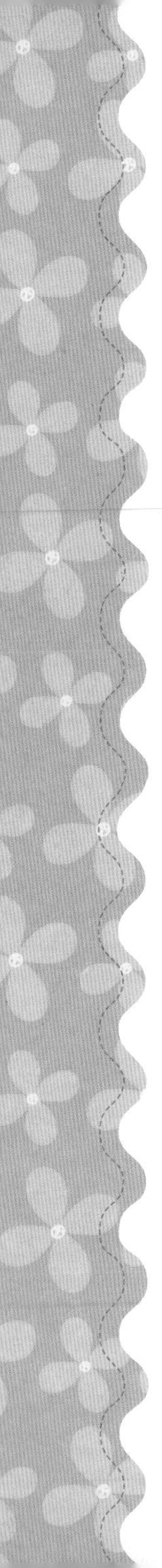

What You Cut

Gray felt
- *2 purse body pieces*
- *2 side pieces*
- *1 bottom piece*

Turquoise felt
- *2 purse body pieces for the lining*
- *2 side pieces*
- *1 bottom piece*
- *1 of each: flowers A and F, and flower center D*

Yellow felt
- *1 of each: flower centers A and H, flower base C, and flower E*

Orange felt
- *1 of each: flowers B, C, and G (use scalloping scissors for flowers C and G if desired)*

Red-orange felt
- *1 of each: flower base B, flower centers C and G, and flower D*

Green felt
- *1 of each: flower base A and flower center B*

Lime felt
- *3 leaves*

Brown felt
- *1 of each: flower center F and flower H*

Plastic canvas
- *2 purse handle supports*
- *1 purse bottom support*

WHAT YOU DO

1 Enlarge the templates on page 125 and cut them out. Cut the fabric as described in the box at left.

2 To make the large flowers A–C, stack the flower center, flower petals, and flower base. Position on the purse front and secure the layers together with decorative stitches as desired.

3 For each small flower, stack the flower center on top of the flower petals and position on the purse back. Use decorative stitches to secure the layers. Stitch on the leaves.

4 Sandwich the plastic canvas handle support between the purse front and the front lining. Pin all three layers together. Embroider blanket stitches around the handle opening. There's no need to catch the handle support in your stitches—it will stay in place. Repeat for the back panel.

5 Place the bottom support piece between the purse bottom and the lining purse bottom. Pin through all of the layers to keep them in place.

figure 1

6 With the turquoise sides together, place the bottom edge of the front panel on one long edge of the tote base (figure 1). Using floss and a blanket stitch, stitch the two pieces together. Repeat for the back panel.

7 Layer one side panel of gray and one side panel of turquoise together. Place on the purse base with the turquoise sides together. Blanket stitch the side to the base. Repeat for the other side. Blanket stitch each side seam and across the purse opening.

GET YOUR BLANKET STITCH ON

This is a great purse for hand sewers. It relies on the blanket stitch (page 30), so make a few practice stitches on a fabric remnant and find your blanket stitch groove.

carry on

Totes for the girl who has everything and needs to bring every bit of it along with her.

the duchess

DESIGNER

NATHALIE MORNU

Whether you're catching the subway or a movie premiere, you'll feel like royalty with this bag on your arm. The bag features rich fabrics, a roomy design, and a decorative appliqué accent with shimmering sequins.

WHAT YOU NEED

Basic Purses Tool Kit (page 11)

½ yard (45.7 cm) of fabric A, a flocked linen for the body of the bag

½ yard (45.7 cm) of fabric B, a cotton fabric for the lining

¼ yard (22.9 cm) of fabric C, a heavy satin fabric for the appliqué and band

7 x 7-inch (17.8 x 17.8 cm) square of fusible interfacing

¼ yard (22.9 cm) of fabric D, a rough-weave fabric for the straps

10 sequins of various sizes

SEAM ALLOWANCE

⅝ inch (1.6 cm) unless otherwise noted

What You Cut

Fabric A

- *two body pieces*

Fabric B

- *two body pieces*

Fabric C

- *one 7 x 7-inch (17.8 x 17.8 cm) piece for the appliqué, and four purse bands*

Fabric D

- *one 5 x 40-inch (12.7 x 100 cm) strip for the strap*

WHAT YOU DO

1 Enlarge the templates on page 121 and cut them out. Cut out the fabric as described in the box at left.

2 Fuse interfacing onto the wrong side of fabric C, and cut out the appliqué from this piece.

3 Pin or baste the appliqué to the lower right corner of the right side of the fabric A purse front, keeping it at least 1 inch (2.5 cm) from the edge. Attach it by topstitching. Sew on the sequins randomly by hand.

4 Make the pleats on all the purse body pieces cut from fabrics A and B and baste them in place. On each piece, sew gathering stitches along the bottom edge between the circles,

THE
EW YOR

following the instructions for Making Gathering Stitches on page 21. Gather so the area measures 4 inches (10.2 cm) long (figure 1). Set aside.

5 Sew one fabric C band to one fabric B lining, positioning the pieces with wrong sides together and notches matching. Stitch. Repeat with another fabric C band and one fabric B lining. Do the same with the fabric A purse body pieces and the remaining fabric C bands. Press and topstitch on the band, ⅛ inch (3 mm) from the seam line (figure 2).

6 Pin the purse front and the purse back with the right sides together. Stitch the sides and bottom of the bag. Clip the curves following the instructions on page 19 and press.

7 Insert the lining following the instructions on page 24 for Lining A using a ⅝-inch seam allowance. Topstitch the bag opening ⅛ inch (3 mm) from the seam.

8 Use the fabric D strip to make one strap, following the instructions on page 26 for Strap B. Cut the strap into two 18-inch (45.7 cm) pieces, and discard the extra. On each piece, turn under the short ends 1 inch (2.5 cm) and press. Topstitch the straps to the band of the bag, 1½ inches (3.8 cm) from the side seams.

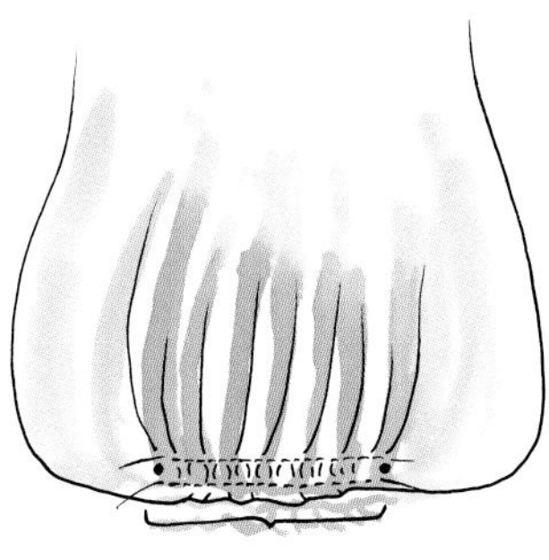

figure 1

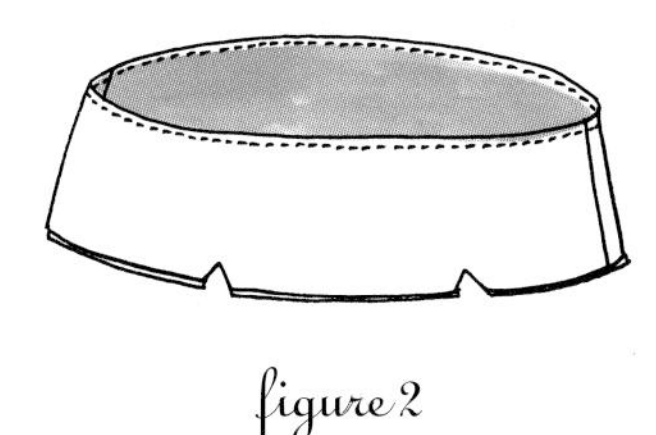

figure 2

STIFF UPPER LIP

Use upholstery fabrics for all but the lining to give this bag the support it needs to hold its shape.

bowled over

Whether you're jetting around the globe or just daydreaming about it, you'll blow everyone away with this ultra-hip bag. The geometric blocks are inspired by video games, and the style is bowling chic for a look that says your game is on.

DESIGNER

RACHEL FIELDS

WHAT YOU NEED

Basic Purses Tool Kit (page 11)

½ yard (30.5 cm) of fabric A, textured cotton

½ yard (30.5 cm) of fabric B, silk

¼ yard (22.9 cm) of fabric C, smooth-weave linen

1 yard (91.4 cm) of fabric D, heavyweight upholstery fabric for the lining

2 yards (1.8 m) of boning

2 yards (1.8 m) of fusible interfacing

Thin foamboard

24-inch (61 cm) zipper

Thread in matching shades

SEAM ALLOWANCE

⅝ inch (1.6 cm) unless otherwise noted

What You Cut

Fabric A
- *4 purse body pieces*
- *2 bottoms*
- *2 sides*

Fabric B
- *4 pocket bands*
- *2 straps*

Fabric C
- *4 pockets*
- *4 purse tops*

Fabric D
- *2 inserts*
- *4 purse body pieces*
- *2 sides*
- *2 straps*

Interfacing
- *2 inserts*
- *2 pocket bands*
- *4 purse body pieces*
- *2 sides*
- *2 straps*

WHAT YOU DO

1 Enlarge the templates on page 129 and cut them out. Cut out the fabric as described in the box on page 92.

2 Cut ¼-inch (6 mm) strips of fabric A and fabric B and arrange them into a patchwork. Sew the pieces together following the instructions on page 20 for Basic Piecing. Make two panels, each 14 x 9 inches (35.6 x 22.9 cm).

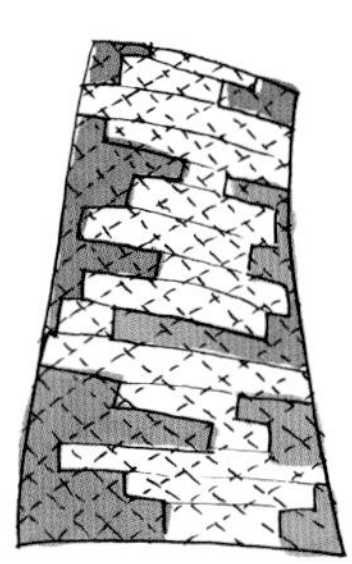

figure 1

3 Cut two insert pieces from the patchwork panels. Fuse the interfacing insert to the wrong side of the insert patchwork panels, following the manufacturer's instructions. Top-stitch both layers in a diamond pattern (figure 1). Repeat for the second patchwork insert.

4 Fuse two interfacing bands to the wrong sides of two fabric B bands. Mark these "pocket band fronts"; mark those without interfacing "pocket band backs."

5 With the right sides together, sew one fabric B front pocket band to the upper edge of the fabric C left pocket front. Clip the seams, following the instructions on page 19, and press. With the right sides together, sew a fabric B back pocket band to the fabric C left pocket back. Sew the left pocket front to the left pocket back at the top of the band with the right sides together. Turn, press, and baste along the lower edge. Repeat with the right pocket pieces.

6 Topstitch a diamond pattern across each pocket, making sure not to sew through the band at the upper edge. Backstitch at the beginning and end of each line of stitching to anchor it.

7 Attach the insert to the fabric A front pieces following the instructions on page 24 for Sewing an Insert. Repeat with the fabric A back pieces. Do the same with the lining front and lining back pieces.

8 Baste the boning along the arch of the purse front. Repeat for the purse back.

9 Baste the front lining to the purse front with wrong sides together. Repeat with the back lining and purse back. Set aside.

10 Turn under ⅝ inch (1.6 cm) of the long side of two top pieces and press. Lay the zipper with the right side facing up. Pin one purse top to the left side of the zipper, aligning the pressed edge with the center of the zipper. Using a zipper foot, sew very close to the zipper teeth. Repeat for the right side.

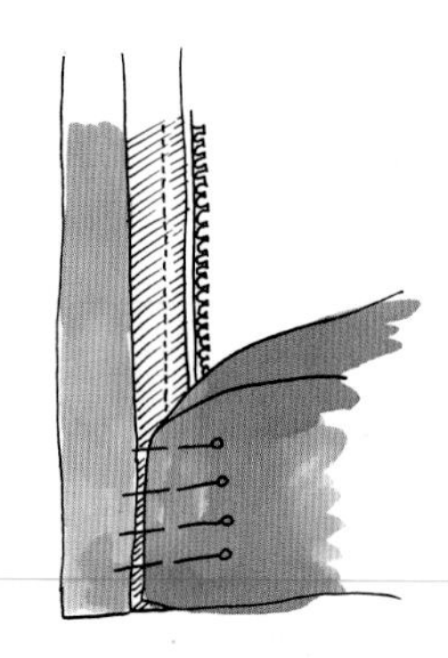

figure 2

11 Turn the sewn zipper panel with the wrong side facing up. Place another top piece with the wrong side facing up over the zipper and align the raw edge with the left side of the zipper tape, over the zipper stitching just sewn. Pin in place. Flip the panel over and topstitch the lining, over the previous zipper stitching line to the right of the zipper teeth (figure 2). Press the fabric underneath back along the hidden seam. Repeat for the other side.

TAG THE BAG

For a project with a lot of fabric and pieces like this one, use a fabric-marking pen to label the wrong side of the pieces when you cut them out. When you're finished sewing, dab the ink with water and it should vanish.

12 Fuse interfacing to the wrong side of the fabric A side piece. Sew the new zipper insert to the top of the side piece with right sides together. Repeat for the other side.

13 Turn under ⅝ inch (1.6 cm) on the top edge of the side piece lining, and pin to the fabric A piece with wrong sides together. Topstitch to secure the lining. Repeat for the other side. Repeat for the zipper insert on the back side, making sure the length of fabric matches the length of the curved purse front/back. ***Note:*** You will not use the full length of the zipper so reinforce the stitching over the zipper where you will cut it off.

14 Fuse interfacing to the wrong side of the fabric B straps. Place one fabric B strap with one fabric D strap, with the right sides together, and sew both long edges. Turn and press. Repeat with the other handle pieces. Topstitch along the long edges.

15 Pin the handle next to the left side of the middle insert with the silk side of the handle facing down. Baste. Baste the other end to the right side of the middle insert for the desired handle length.

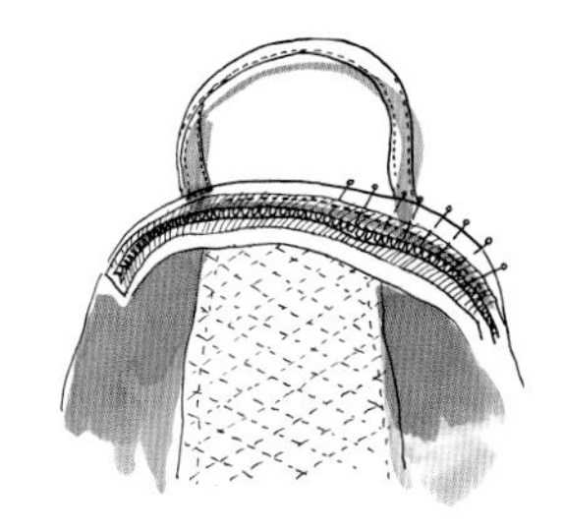

figure 3

16 Sew the purse front to the zipper insert, making sure to reinforce the stitching at the handles (figure 3). Turn, clip, and press. Repeat on the back of the bag.

17 Baste the purse bottom pieces, wrong sides together along three sides, leaving one short side open. Cut a piece of the foamboard to fit within the bottom casing. Insert it and baste the remaining edge, enclosing the board.

18 Turn the entire top of the bag inside out, leaving the zipper open, and stitch the purse bottom onto the purse body. Turn sharply at the corners and take your time to ensure smooth seams. Sew a second time over the seam to reinforce it.

yo, chica

Spice up some basic corduroy with bright yo-yos and you'll be ready to strut your stuff. A magnetic closure keeps everything secure, and there's even a special little pocket inside to hold your phone.

WHAT YOU NEED

Basic Purses Tool Kit (page 11)
1/3 yard (30.5 cm) of corduroy
Scrap of fabric for the phone pocket
1/3 yard (30.5 cm) of cotton print fabric for the lining and the yo-yos
Magnetic clasp
Plastic canvas, 3/4 x 9 inches (9.5 x 22.9 cm)

SEAM ALLOWANCE

1/4 inch (6 mm) unless otherwise noted

DESIGNER

WENDI GRATZ

What You Cut

Scrap fabric

- *2 rectangles for the pocket, each 4 1/4 x 3 1/4 inches (10.8 x 8.3 cm)*

Cotton print

- *2 rectangles for the lining, each 14 x 10 1/2 inches (35.6 x 26.7 cm)*
- *1 circle 3 inches (7.6 cm) in diameter*
- *1 circle 3 1/2 inches (8.9 cm) in diameter*
- *1 circle 5 1/4 inches (13.3 cm) in diameter*

Corduroy

- *2 rectangles for the body of the purse, each 14 x 10 1/2 inches (35.6 x 26.7 cm)*
- *2 rectangles for the purse handles, each 4 1/2 x 15 inches (11.4 x 38.1 cm)*

WHAT YOU DO

1 Cut the fabric as described in the box at left. Pin the two corduroy rectangles with the right sides together and stitch around the sides and bottom. Box the bottom of the purse by flattening one bottom corner into a point, lining up the side and bottom seams. Stitch 2 inches (5.1 cm) from the point, perpendicular to the seam (figure 1). Repeat with the second corner.

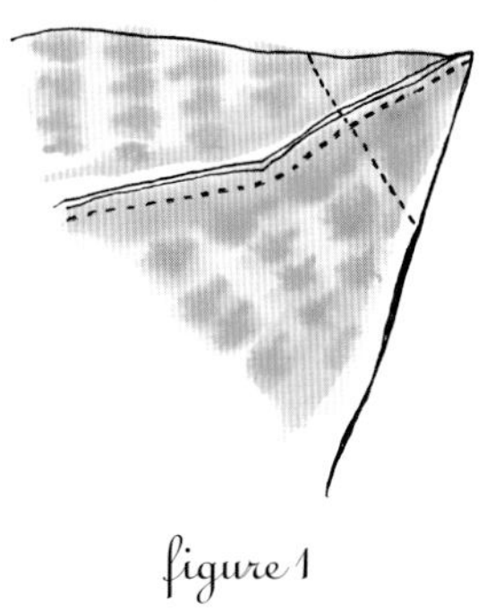

figure 1

2 To sew the purse handles, follow the instructions on page 26 for Strap B. Mark the center of the purse front. Pin one end of a strap so that the inner edge is 2 1/4 inches (5.7 cm) from the center marking. Repeat with the other end. Pin the other strap on the purse back in the same fashion. Baste.

3 Make the cell phone pocket for the lining following the instructions on page 25. Sew the pocket in place on one of the lining panels.

4 Make the lining following the instructions on page 24, boxing the bottom as described in step 1 (page 99). Before attaching the lining to the purse, attach the magnetic clasp to the lining, following the manufacturer's instructions. Then, complete the instructions for lining the bag and topstitch around the opening ¼ inch (6 mm) from the edge.

5 Follow the instructions on page 29 to make the yo-yos from the three circles of print fabric. Position the yo-yos on the bag and stitch them down by hand.

6 Place the plastic canvas in the bottom of the bag. Sew it down by hand, or leave the plastic loose if you want to be able to remove it for washing the bag.

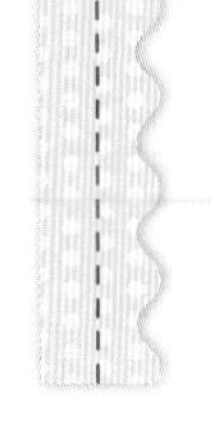

DON'T BUDGE

Use plenty of pins when positioning corduroy for sewing—this fabric shifts a lot.

beach baby

Life's a beach when you tote this bag. Not only is it spacious enough for a towel and tanning lotion, but it's also reversible so you can hit the boardwalk in style.

DESIGNER

NATHALIE MORNU

WHAT YOU NEED

Basic Purses Tool Kit
(page 11)

½ yard (45.7 cm) of fabric A, a cotton print for the bag

½ yard (45.7 cm) of fabric B, a complementary cotton print for the lining

¼ yard (22.9 cm) of fabric C, a complementary cotton stripe for the pockets

¼ yard (22.9 cm) of fabric D, a solid cotton for the bias tape and straps

SEAM ALLOWANCE

½ inch (1.3 cm) unless otherwise noted

WHAT YOU DO

1 Enlarge the templates on page 120 and cut them out.

2 Make a ¼-inch (6 mm) hem along the top of each pocket. Press the lower edge of each pocket under ½ inch (1.3 cm), clipping the curves carefully.

3 Pin a pocket, right side out, to the right side of one of the bag pieces cut from fabric A. Pin the other pocket, right side out, to the right side of one of the bag pieces cut out of fabric B. Topstitch along the sides and bottom edge of each pocket to attach it to the bag, leaving the top open. Topstitch down the center of the pocket to create two compartments.

BRAID IS AN AID

If you don't feel like making bias tape, you can use purchased fold-over braid, sold in the trimmings department of fabric stores, to bind the bag and make the straps.

4 Pin together both bag pieces cut from fabric A, with the right sides facing and matching all edges. Stitch along the curved bottom edge. Press the seam open. Repeat with fabric B to make the lining.

5 Add the lining following the instructions on page 24 for Lining A.

6 Using fabric D, make a strip of double-fold bias tape 3 inches (7.6 cm) wide and 60 inches (150 cm) long, following the instructions on page 23.

7 On each side of the bag, bind the edge of the central arc—the one directly above the pocket—with a piece of bias tape, following the directions for Binding a Seam on page 22. Trim off any excess tape.

8 The straps are both made from a single piece of bias tape which also encases the entire edge of the bag, and both the straps and the encasing are stitched in the same pass. Proceed as follows: Begin pinning the bias tape at one side seam, turning under the end so that when you stitch it down later, you'll get a clean seam (page 22, figure 7). Work your way around the bag, pinning the tape until you reach one of the arcs where tape was applied in step 6. Leave 20 inches (50.8 cm) of bias tape loose to use as a strap. Resume binding on the other side of the arc, and continue pinning the bias tape around the edge of the bag until you reach the other arc. Again leave 20 inches (50.8 cm) of loose tape for the other strap, and resume binding on the other side of the arc. Continue pinning the bias tape around the edge until you reach the starting point. Cut off any excess tape and stitch the entire piece of tape. This not only attaches it to the bag, but also closes the bias tape that makes up the straps.

patch madame

One look at this patchwork bag and you'll be wanting one of your own. Dig through your fabrics and get inspired. Soon you'll be patchin' with panache.

DESIGNER

ELIZABETH SEARLE

WHAT YOU NEED

Basic Purses Tool Kit (page 11)

⅝ yard (57.2 cm) of Fabric A, a colored print or solid for the lining

⅝ yard (57.2 cm) of Fabric B, muslin for the body of the bag

½ yard (45.7 cm) of Fabric C, a contrasting fabric for the binding

⅝ yard (57.2 cm) of quilt batting

Fabric scraps to cover the batting

Several spools of thread

Washable glue sticks

SEAM ALLOWANCE

½ inch (1.3 cm) unless otherwise noted

WHAT YOU DO

MAKE THE PATCHWORK

1 Cut the fabric as described in the box at right. Cover the fabric B muslin with varying sizes of fabric scraps. Lightly glue the pieces in place with a glue stick, overlapping the edges by at least ½ inch (1.3 cm).

2 Turn the patchwork face down. Place the batting on the wrong side of fabric B and pin securely to hold it in place.

3 Quilt the layers together densely in a back and forth zigzag pattern, making sure to secure each piece of fabric (figure 1). Machine wash and dry the patchwork to gently fray the fabric edges and create texture.

4 After drying, fill any holes or gaps with fabric scraps. Glue the new pieces into place and quilt over them.

5 Make a pattern for the purse body by drawing an 11-inch (27.9 cm) circle on scrap paper. Draw a small curve (or trace around a bowl) that intersects the top of the circle to make the line that will be the top of the bag. Cut out the pattern.

What You Cut

Fabric A
- *2 purse pieces on the fold*

Fabric B
- *2 lining pieces on the fold*

Fabric C
- *2 pockets on the fold*

6 Cut two pieces of the purse body from the quilted material. Cut two strips of the quilted fabric 2½ to 3 inches (6.4 to 7.6 cm) wide. Measure the circumference of the purse body piece and add the desired length for the purse strap. (The strap shown here is 35 inches [87.5 cm] long.) Cut the quilted strips to this length. Join the ends of the strip with a wide zigzag stitch. Cover with a small fabric scrap, and quilt it in place.

7 Cut the remaining strip 10 inches (25.4 cm) long, trimming one end into a curve. Set it aside to use later as the purse flap.

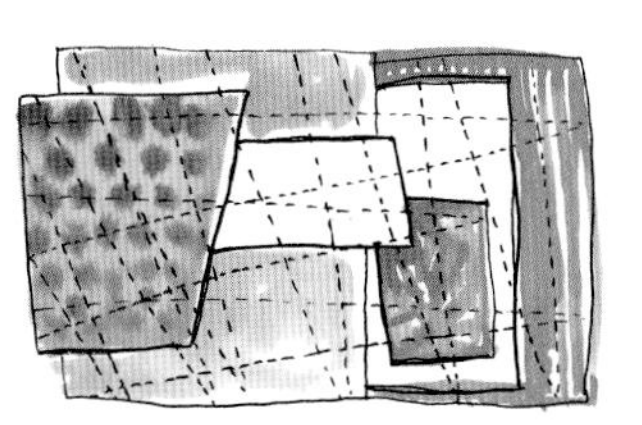

figure 1

8 Pin the long strap around the curved side of the purse back and baste. Repeat for the purse front.

9 Follow the instructions on page 23 to make bias tape from fabric C, starting with strips 3 inches (7.6 cm) wide and long enough to go around each side of the strap. Cut a short piece of bias tape to go around the flap edges.

10 Follow the instructions on page 22 for Binding a Circumference to bind both long edges of the strap and side seam all the way around.

11 Follow the instructions on page 22 for Binding a Seam to attach the bias tape to the purse flap, encasing the sides and curved edge of the flap, leaving the short straight edge unbound.

12 Position the flap on the center of the purse back and baste it in place. Cover it with a fabric scrap, and quilt to secure the flap to the bag (figure 2). Embellish the flap with a bead or flower to add weight and hold it in place.

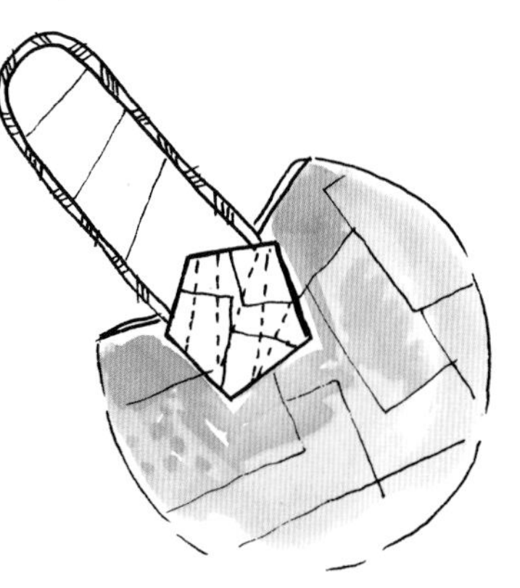

figure 2

BATTING 1000

Instead of using a lofty padding, sandwich flannel or muslin between the quilted layers to give the bag a trimmer appearance.

dotted bliss

Here's the everyday bag you've been searching for. Not only is it the perfect size, but with denim fabric and a fetching bow, you can dress your look up or down. A contrasting strip of fabric at the bottom of the bag adds extra sass.

WHAT YOU NEED

Basic Purses Tool Kit (page 11)

½ yard (45.7 cm) of fabric A, a lightweight denim

⅜ yard (34.3 cm) of fabric B, a large polka-dot print cotton for the handle, bow, and sides

½ yard (45.7 cm) of fabric C, a small polka-dot print cotton for the lining

SEAM ALLOWANCE

½ inch (1.3 cm) unless otherwise noted

DESIGNER

MORGAN MOORE

What You Cut

Fabric A

- *two purse pieces*

Fabric B

- *two sides, one band, one handle, and two bows*

Fabric C

- *two purse pieces for the lining and one bottom*

WHAT YOU DO

1 Enlarge the templates on pages 126 and 127 and cut them out. Cut the fabric as described in the box at left.

2 Transfer the dart markings for the purse to the fabric A purse front and back. Pin and baste the darts in place.

3 Fuse the interfacing pieces for the bottom and sides to the wrong sides of fabric B side pieces and fabric C bottom.

4 With the right sides together, align one short edge of the fabric C bottom with one fabric B side, and stitch. Align the other short edge of the fabric C bottom with the other fabric B side, and stitch. Press the seams. This piece is the side insert.

DARTING AROUND

You don't need to make darts in the lining even though you do stitch them on the exterior fabric. That's because the narrow band of material showing at the side seam makes up the difference.

5 With the right sides together, position the side insert onto the fabric A purse front. Center the bottom panel of the side insert at the bottom of the bag and pin from the bottom up each side. Let the excess side fabric extend off the top of the bag. Stitch the seam. Repeat with the fabric A purse back. Turn the bag right side out and press.

6 With the right sides together, stitch the fabric C lining pieces together at the sides and bottom, leaving the top open. Tuck the lining inside the bag. Match the top edges all around and baste.

7 Fold the fabric B band in half lengthwise with wrong sides together and press to crease. Turn under ¼ inch (6 mm) on the long edges and press. Open the center fold and pin the short edges of the band together, leaving the ¼-inch (6 mm) folds in place. Stitch along the short edge of the band. Refold the band along its center crease with right sides out.

8 Slip the open edge of the band ¼ inch (6 mm) over the top basted edge of the bag, matching the band seam to one side seam. Pin the band in place and stitch close to the edge, attaching the band to the bag.

9 Sew the fabric B strap following the instructions on page 26 for Strap A. Stop after step 1 and tuck in the raw edges on the short end and stitch. Repeat Strap A, step 1, with the pieces for the bow.

10 Place the sewn strips for the bow on top of each other and stitch one short edge. Place the stitched edge along the bottom of the band over the side seam and stitch in place (figure 1).

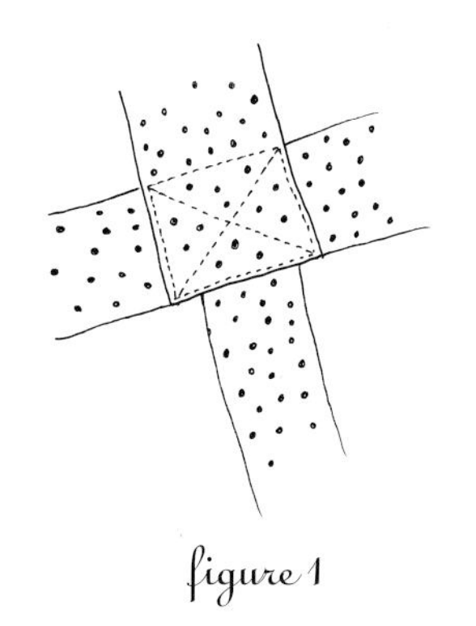

figure 1

11 Pin each end of the purse handle to the lower seam of the band on the side seams of the bag. Stitch an X to secure each end of the handle in place (figure 2).

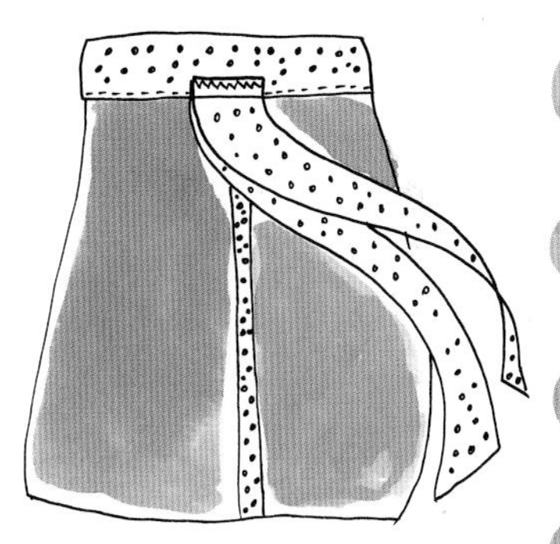

figure 2

ruffled delight

Raid your fabric stash for all your favorite scraps and turn them into layer upon layer of flirty-girl ruffles. Once you're done, you'll want to shake it, baby, shake it!

DESIGNER

DORIE BLAISDELL SCHWARZ

WHAT YOU NEED

Basic Purses Tool Kit (page 11)
½ yard (45.7 cm) of heavyweight fusible interfacing
½ yard (45.7 cm) of fabric A, a muslin for the purse body
½ yard (45.7 cm) of fabric B, a cotton print or solid for the lining
¼ yard (45.7 cm) of fabric C, a heavy cotton for the straps
⅜-yard (34.3 cm) of eight different cotton prints for ruffles
1 package of ½-inch (1.3 cm) double-fold bias tape
Thread to match bias tape and handle fabrics
40 plastic pony beads
Scraps of material for yo-yos
3 buttons
10 inches (25.4 cm) of rickrack

SEAM ALLOWANCE

½ inch (1.3 cm) unless otherwise noted

What You Cut

Fabric A
- *Refer to step 2*

Fabric B
- *2 sides and cut one bottom for the purse lining*

Interfacing
- *2 sides*
- *1 bottom*

Fabrics for ruffles
- *1 strip of each fabric, 5 x 65 inches (12.7 x 165.1 cm). You'll have to piece them to achieve this*

Fabric C
- *2 strips for the handles, each 1¾ x 50 inches (1.9 x 125 cm)*

WHAT YOU DO

1 Enlarge the templates on page 123 and cut them out. Cut the fabric as described in the box above.

2 Fuse the interfacing side pieces to the wrong side of the uncut fabric A muslin and cut out. Fuse the interfacing bottom to the wrong side of an uncut piece of fabric B lining and cut out. This is the purse bottom for the outside of the bag.

3 Sew the fabric A muslin side pieces together to make the body of the bag, and turn right side out.

4 Fold one ruffle strip lengthwise into a 2½-inch (6.4 cm) strip with wrong sides together and press. Sew gathering stitches, following the instructions on page 21, ¼ inch (6 mm) from the long edge of each ruffle piece. Gather each ruffle so it's slightly longer than the circumference of the bag.

5 Mark the bag for placing the ruffles by drawing horizontal lines around the muslin in 1½-inch (3.8 cm) increments from the bottom up.

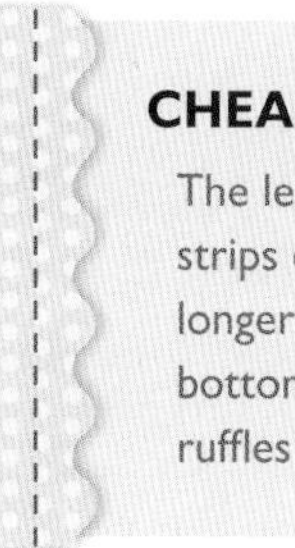

CHEAP FRILLS

The length of your ruffle strips can vary slightly—use longer ruffles for the wide bottom of the bag and shorter ruffles toward the top.

figure 1

6 Pin the bottom ruffle in place, aligning the raw edge of the ruffle with the bottom line. Turn under the narrow end of the ruffle strip and pin the strip around the bag. Trim any extra ruffle, leaving 1 inch (2.5 cm) of overlap. Fold the overlap allowance over twice to hide the raw edge, and pin. Stitch the ruffle to the bag, sewing ¼ inch (6 mm) from the raw edge of the ruffle. Repeat with the next five ruffles (figure 1). As the layers become thick and harder to manage, just feed the fabric little by little, pushing the gathers aside and stopping to re-align the fabric as needed.

7 Cut the remaining two ruffles in half. Attach one half to the purse front and one half to the purse back, turning under the ends of each piece and pinning them to the wrong side of the slanted opening of the bag. Repeat with the last two sections.

8 Turn the bag inside out, and sew on the fabric B purse bottom with the right sides together. Clip the curves following the instructions on page 19 and turn right side out.

9 With the right sides facing, sew the fabric B lining side pieces together and press the seams open. Sew on the fabric B lining bottom and clip the curves, trimming the seam to ¼ inch (6 mm). Place the lining into the purse, wrong sides facing, smoothing out the seams for a good fit.

10 Bind the top edge of the bag and miter the corners following the instructions on page 22 for Binding a Seam with Corners. The top of the side seams will still be exposed after binding.

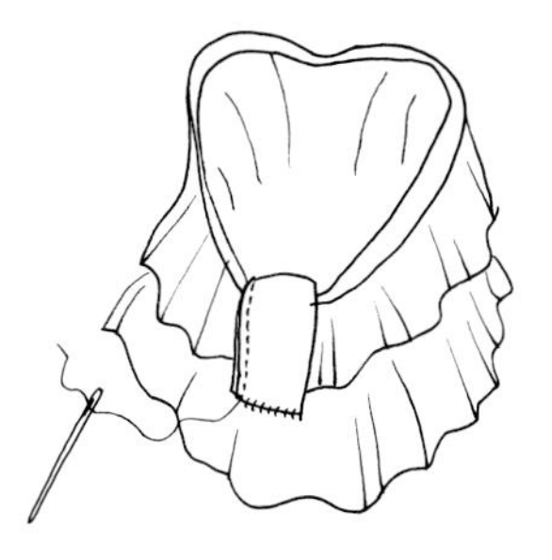

figure 2

11 Cut a short piece of bias tape 3 inches (7.6 cm) long for each side of the bag. Fold the tape in half lengthwise, tuck under the ends, and stitch. Pin the tape over the side seam with one end outside the bag and one end inside the bag, and hand stitch in place (figure 2).

12 Make the straps with the fabric C pieces following the instructions on page 26 for Strap A, using a ¼-inch (6 mm) seam allowance. Tie a knot in the sewn strap about 4 inches (10.2 cm) from one end. Slide a pony bead into the tube, down to the knot. Tie another knot above the bead and slide another bead into the tube. Repeat for the length of the tube until 4 inches (10.2 cm) remain, or the handle is the desired length. Repeat with the other strap.

13 Make a ½-inch (1.3 cm) buttonhole by hand with the buttonhole stitch (page 30) or by machine. Place it 1½ inches (3.8 cm) in from each side seam on the front and back of the bag, just under the edge of the bias tape. Pull the ends of the handles through each of the buttonholes from the outside of the purse.

14 Fold the 4-inch (10.2 cm) ends of the handle to the back, forming a loop through the buttonhole. Tuck in the raw edges, and stitch the loop closed.

15 Make three yo-yos of varying sizes following the instructions on page 29, using scraps of material. Stitch the yo-yos to a top corner of the bag, and add buttons to the yo-yo centers. Fold the rickrack in half and stitch onto the bag beside the yo-yos.

HORSIN' AROUND

Plastic pony beads are used inside the straps of this bag for a nicely weighted, easy-to-grasp handle, so be sure to choose a bead color that won't show through your handle fabric.

checkmate

It's time for some plaid-itude, darling, and this bag's got it all. From the faux tortoise shell handles to the sweet ribbon trim, this bag is super savvy. Which may be why, as soon as your friends see it, they'll wonder if you got a promotion.

WHAT YOU NEED

- **Basic Purses Tool Kit** (page 11)
- ¼ yard (22.9 cm) of thick, fusible web
- ¼ yard (22.9 cm) of plaid fabric A
- ¼ yard (22.9 cm) of fabric B, a print cotton
- ¼ yard (22.9 cm) of fabric C, for the lining
- Invisible thread
- 1 yard (91.4 cm) of 1½-inch (3.8 cm) plaid ribbon
- 30 inches (75 cm) of 5/8-inch (1.6 cm) grosgrain ribbon
- 2 plastic purse handles

SEAM ALLOWANCE

½ inch (1.3 cm) unless otherwise noted

WHAT YOU DO

1 Enlarge the templates on page 128 and cut them out.

Cut the pieces from the fabric as described in the box below.

What You Cut

Fabric A
- *2 purse body pieces*

Fabric B
- *4 pockets*

Fabric C
- *2 purse body pieces*

Fusible web
- *2 purse body pieces*

2 Apply the fusible web to the wrong side of the fabric A purse body pieces, following the manufacturer's instructions. Machine baste close to the edge to be sure it stays in place.

3 Make the pockets following the instructions on page 25, using a ¼-inch (6 mm) seam allowance. Leave the gap at the top of the pocket instead of the bottom.

4 Transfer the markings from the template onto the pocket fronts. Fold the pockets to the center on the fold lines and stitch in place (figure 1). Line up the grosgrain ribbon across the top of the pocket, front and back, encasing the raw edge. Machine stitch it in place.

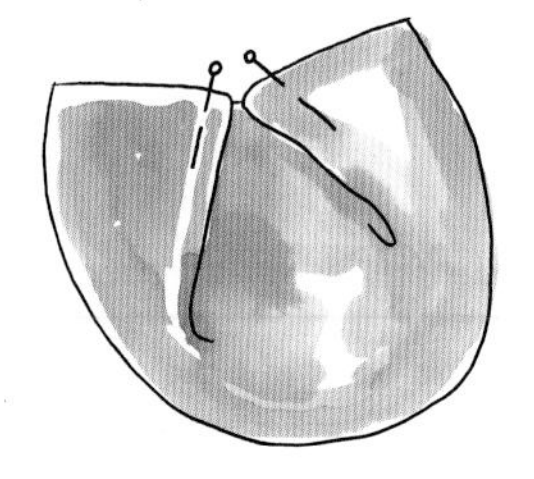

figure 1

DESIGNER

JOAN K. MORRIS

5 Place the pockets in position on one of the Fabric A purse body pieces. To add volume, pin the top corners of each pocket 3 inches (7.6 cm) apart from one another, and pin the bottom of the pocket so that it gathers and puckers. Using invisible thread in the top of the sewing machine and regular thread in the bobbin, edgestitch the pocket in place.

6 Place the fabric A body pieces with the right sides together, and stitch around the sides and the bottom, starting and ending the seam 1½ inches (3.8 cm) from the top edge. Clip the curves and turn the bag right side out.

7 Cut four 6-inch (15.2 cm) pieces of grosgrain ribbon. Fold one in half around the purse handle and use a zipper foot to stitch across the ribbon, as close as possible to the handle (figure 2). Repeat to attach two pieces of ribbon to each handle.

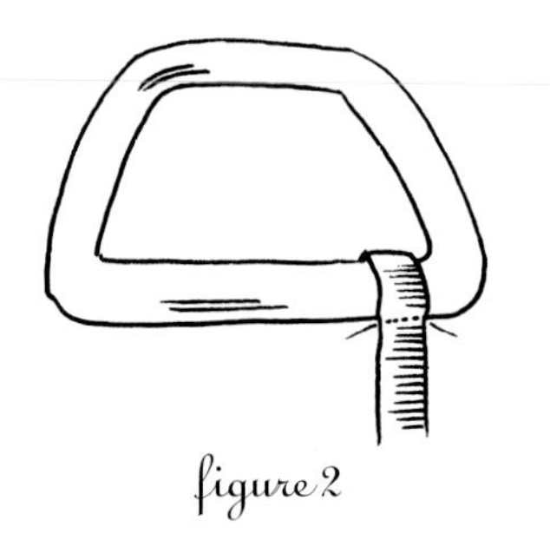

figure 2

8 Place one handle on the front of the bag with the raw edge of the ribbon tabs along the top of the bag. Machine stitch the ribbons in place. Repeat with the other handle on the back of the bag.

9 Attach the lining following the instructions on page 24, using a zipper foot to stitch a ½-inch (1.3 cm) seam, and stitching 1½ inches (3.8 cm) down the side to the side seam.

10 Cut two pieces of the plaid ribbon to fit around the top of the purse, under the handles, with enough to turn under the ends. Pin in place. Hand stitch the tops of the ribbons together using invisible thread, and stitch the bottom of the ribbons to the purse, being sure to secure the folded ribbon end.

OUTTA SIGHT

Sew the grosgrain ribbon in place with invisible thread for a crafty, stitch-free look.

inside out

Ready to throw in the towel? Make a bag instead! Two different colored dishtowels create a reversible bag you can turn inside out whenever the mood strikes.

DESIGNER

REBEKA LAMBERT

WHAT YOU NEED

- **Basic Purses Tool Kit** (page 11)
- 2 kitchen towels of the same size, in complementary colors
- 2 brown plastic purse handles, 8-inch (20.3 cm) rings
- 8-inch (20.3 cm) saucer for tracing

WHAT YOU DO

1 Remove the tags from the kitchen towels and press to remove any wrinkles.

2 Fold one towel in half, with the right sides together, matching up the short edges. Using a plate or saucer, trace a rounded edge onto each corner of the fabric on the fold. Repeat for the other towel.

3 Sew the sides of the folded towel from the 6½-inch (16.5 cm) mark to the rounded marking and then sew along the marking. Reinforce the seam when you stitch following the method on page 18, step 4. Repeat with the other towel.

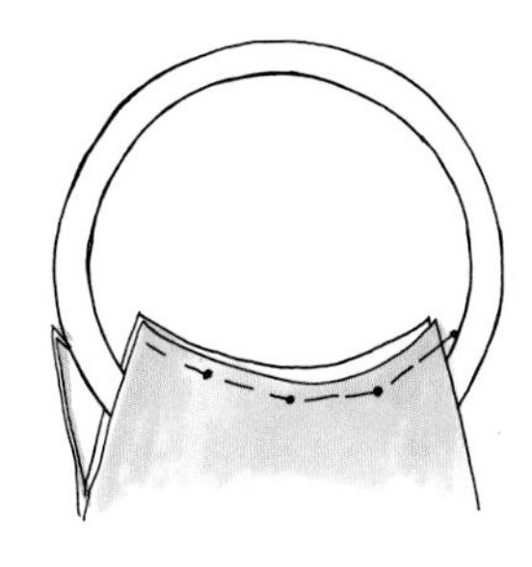

figure 1

4 Insert one bag body into the other with the right sides together. Tuck one purse handle between the purse layers. Pin and stitch the top seam between the inside edges of the handle (figure 1), reinforcing it for durability. Repeat for the other handle. This will leave a gap on each side of each handle. Turn the purse right side out.

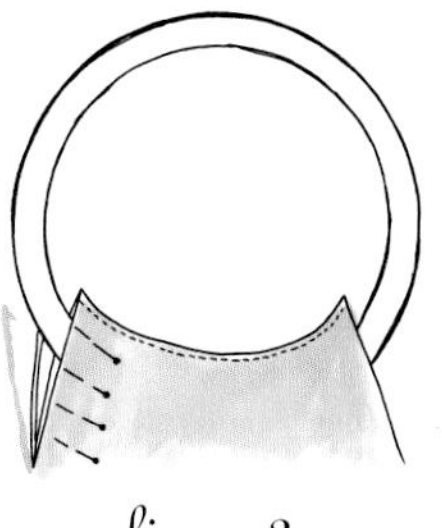

figure 2

5 Working around the handle, sew the purse layers along the side opening, starting 2 inches (5.1 cm) from the handle (figure 2) . Stitch down to the purse's side seam, leave the needle down in the fabric, pivot, and sew across the side seam. Pivot again and sew up the other side, stopping about 2 inches (5.1 cm) from the handle. Repeat on the other side.

6 Starting at the same point, 2 inches (5.1 cm) below the handle, sew a line of stitching for the casing below the purse handle. Stitch slowly, pushing the volume of fabric out of the way as you sew, and reinforcing the stitches at the start and finish. Repeat for the other handle.

templates

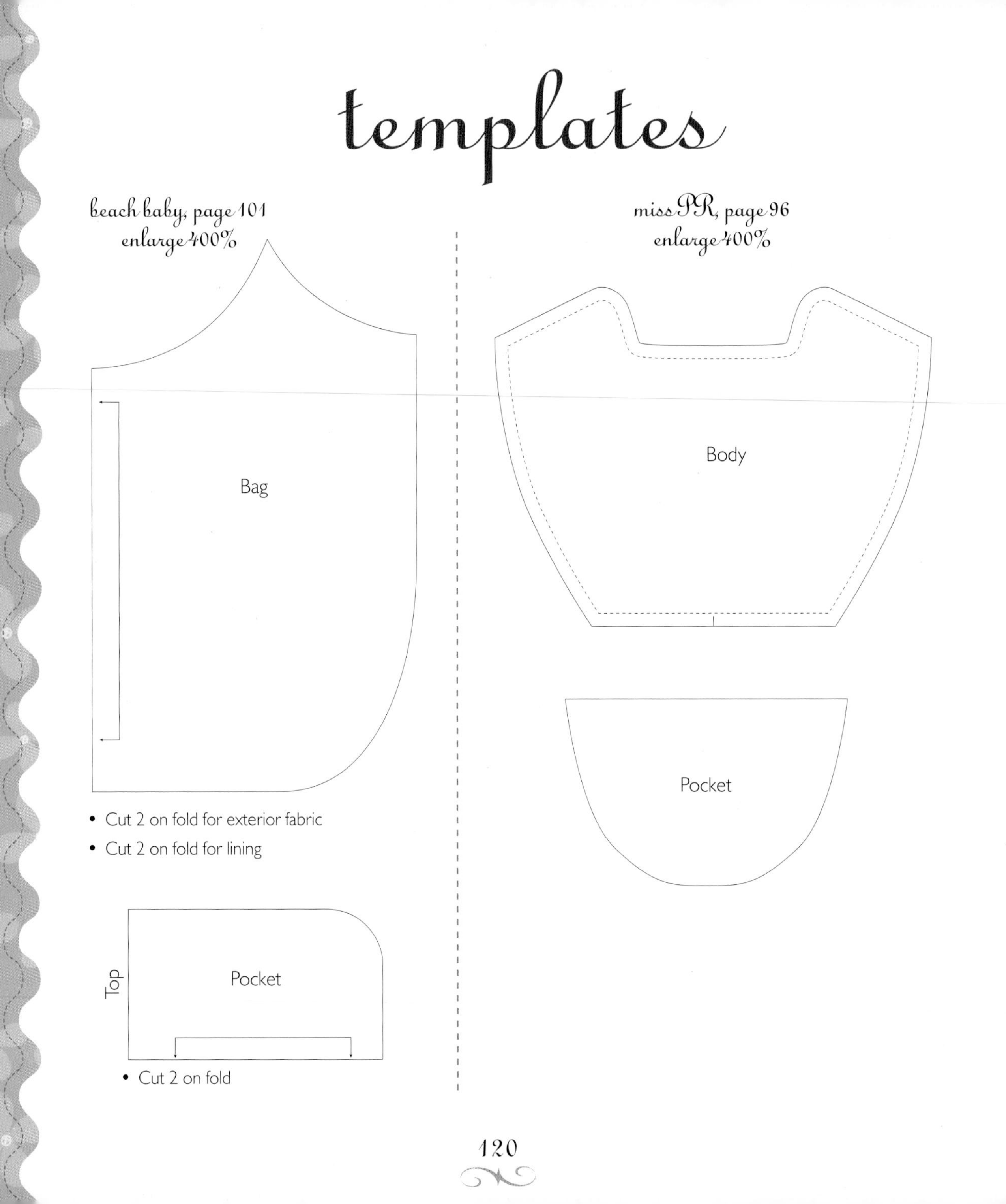
beach baby, page 101
enlarge 400%
Bag
Cut 2 on fold for exterior fabric
Cut 2 on fold for lining
Top
Pocket
Cut 2 on fold
miss PR, page 96
enlarge 400%
Body
Pocket

the duchess, page 88
enlarge 400%

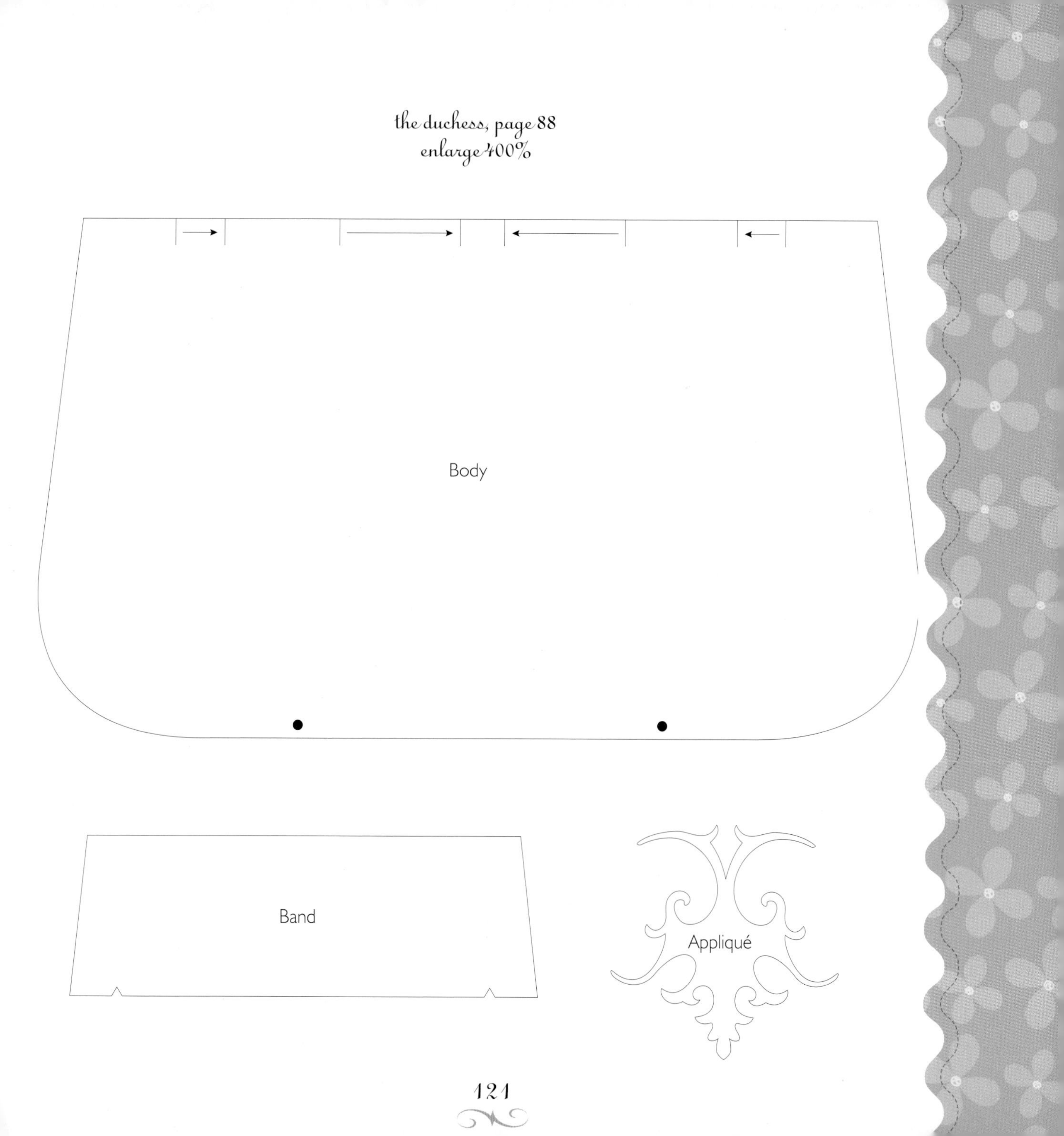

tokyo rose, page 34
enlarge 200%

happy village, page 38
enlarge 200%

ruffled delight, page 110
enlarge 400%

button clutchin', page 58
enlarge 400%

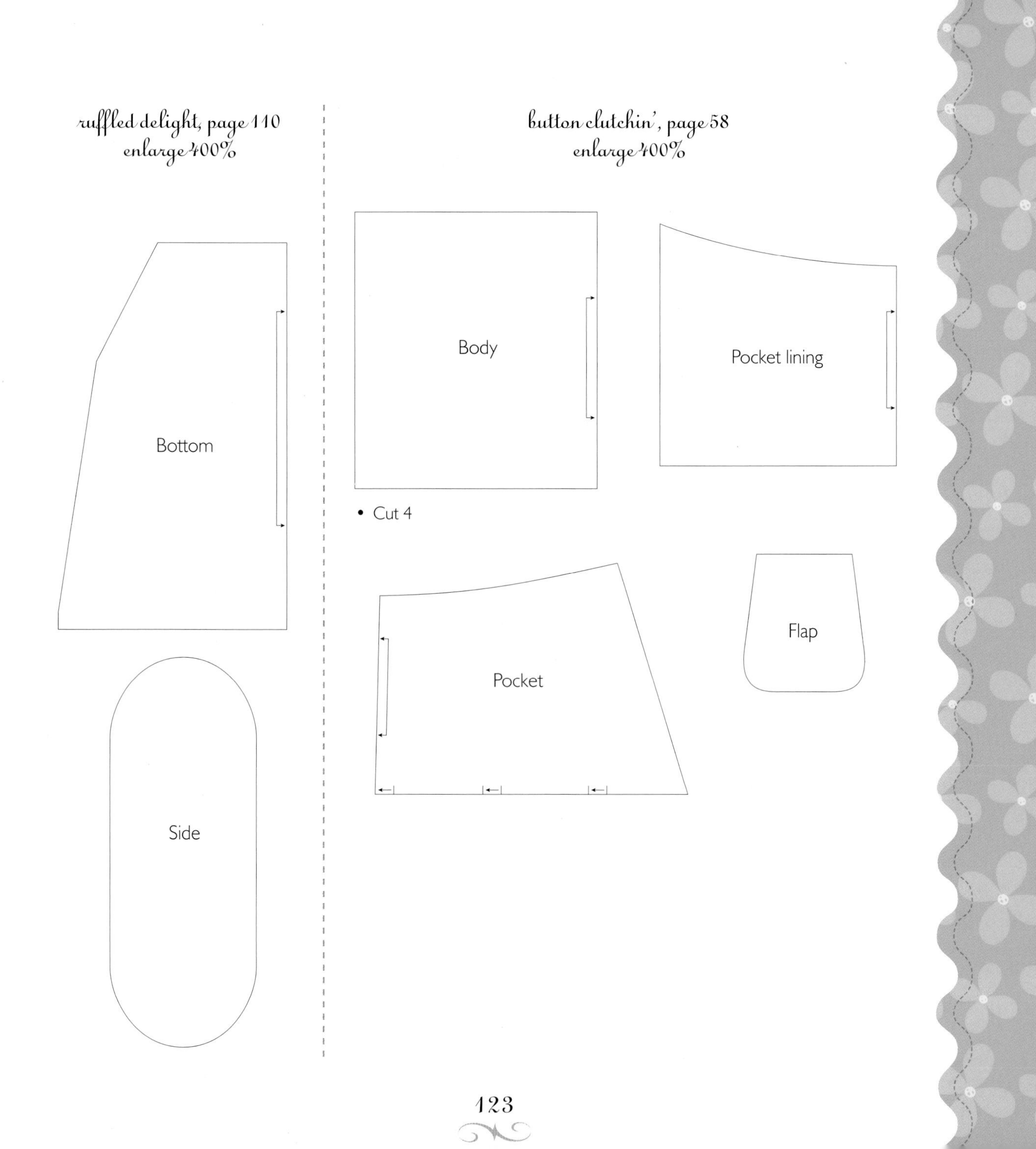

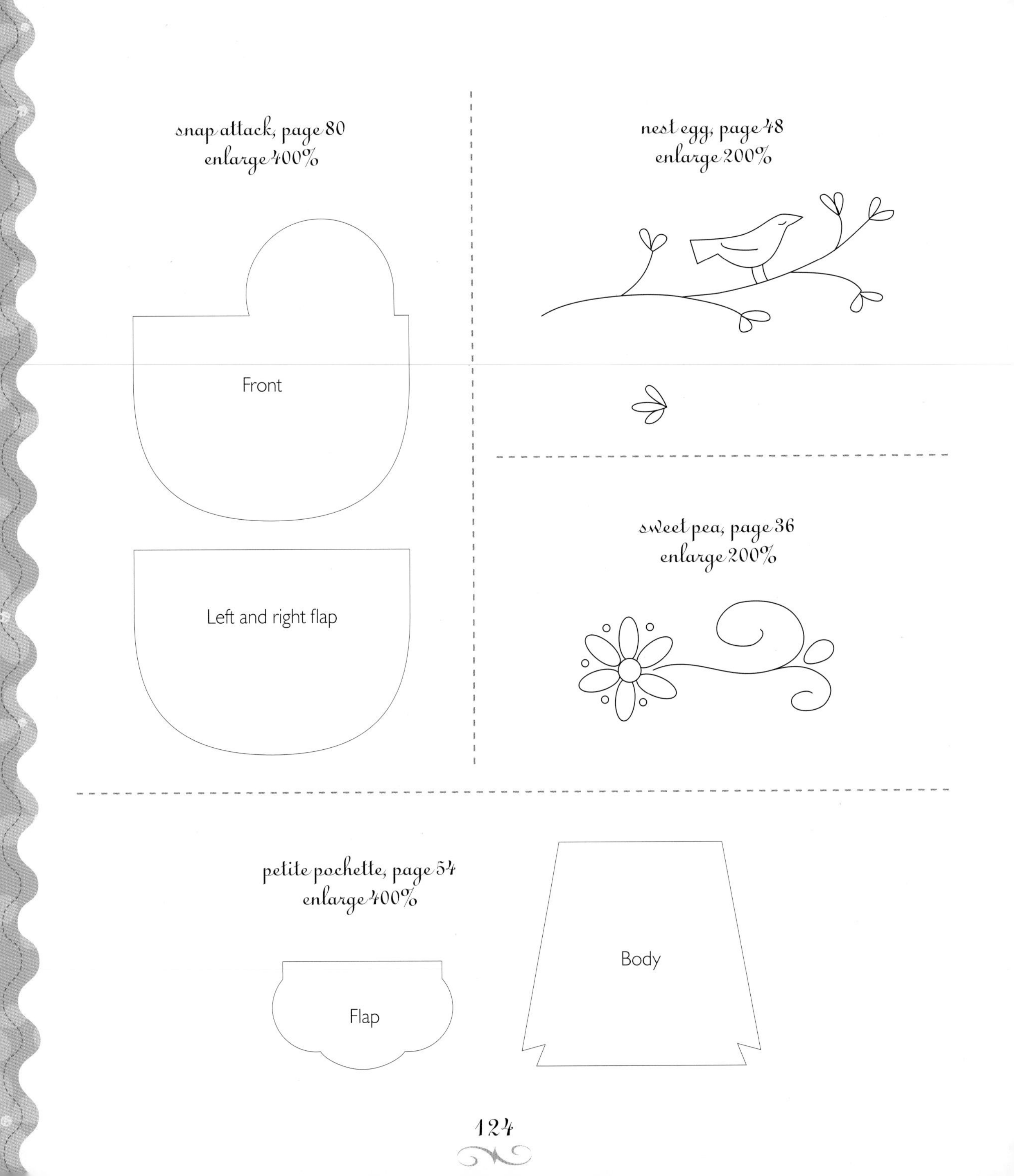
snap attack, page 80
enlarge 400%
Front
Left and right flap
nest egg, page 48
enlarge 200%
sweet pea, page 36
enlarge 200%
petite pochette, page 54
enlarge 400%
Body
Flap

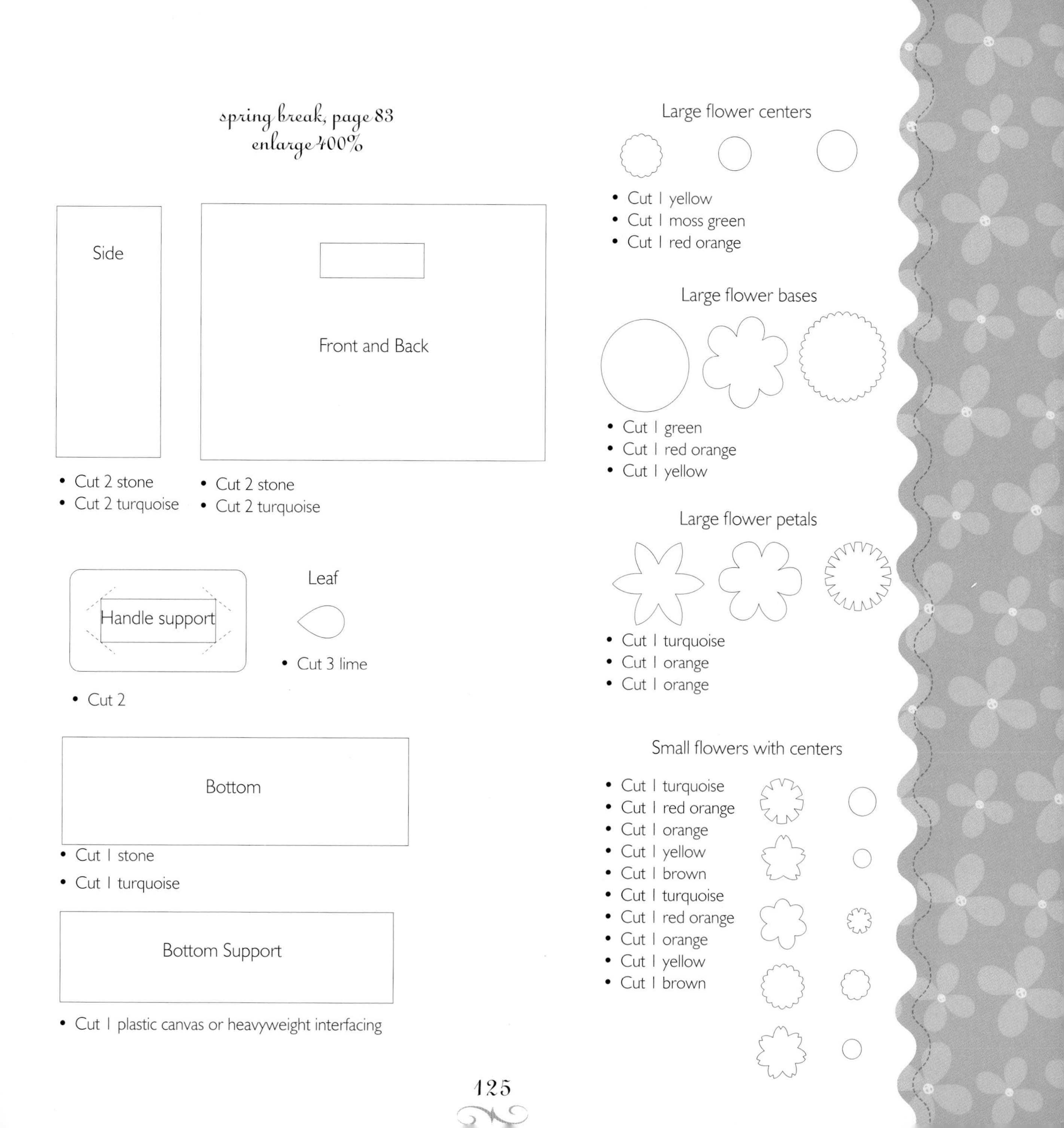
spring break, page 83
enlarge 400%
Side
Front and Back
• Cut 2 stone
• Cut 2 turquoise
• Cut 2 stone
• Cut 2 turquoise
Handle support
• Cut 2
Leaf
• Cut 3 lime
Bottom
• Cut 1 stone
• Cut 1 turquoise
Bottom Support
• Cut 1 plastic canvas or heavyweight interfacing
Large flower centers
• Cut 1 yellow
• Cut 1 moss green
• Cut 1 red orange
Large flower bases
• Cut 1 green
• Cut 1 red orange
• Cut 1 yellow
Large flower petals
• Cut 1 turquoise
• Cut 1 orange
• Cut 1 orange
Small flowers with centers
• Cut 1 turquoise
• Cut 1 red orange
• Cut 1 orange
• Cut 1 yellow
• Cut 1 brown
• Cut 1 turquoise
• Cut 1 red orange
• Cut 1 orange
• Cut 1 yellow
• Cut 1 brown

dotted bliss, page 107, enlarge 500%
(pattern piece for bow appears on next page)

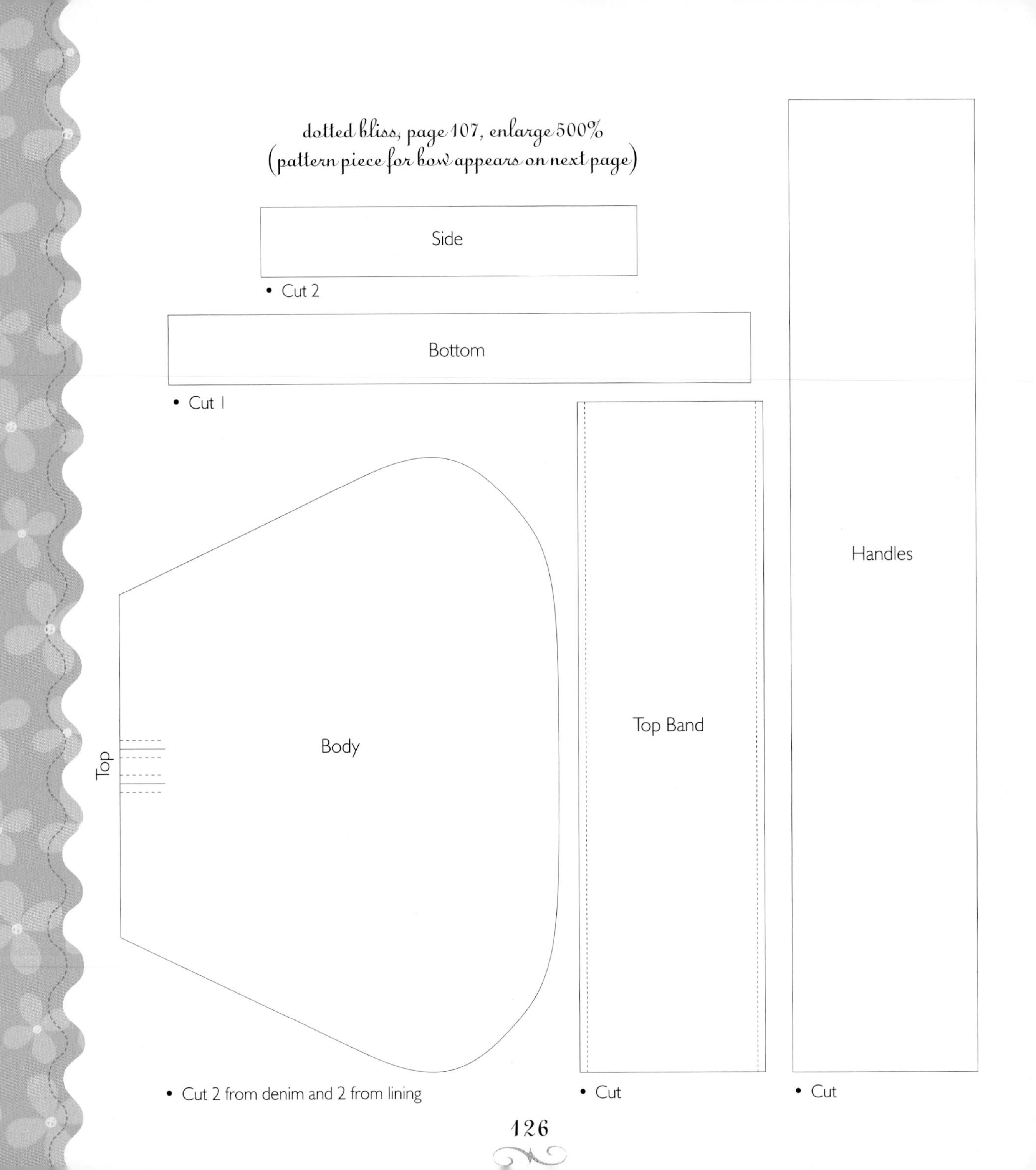

bow for dotted bliss, page 107, enlarge 500%

tea bag, page 50 enlarge 400%

hand bag, page 68 enlarge 400%

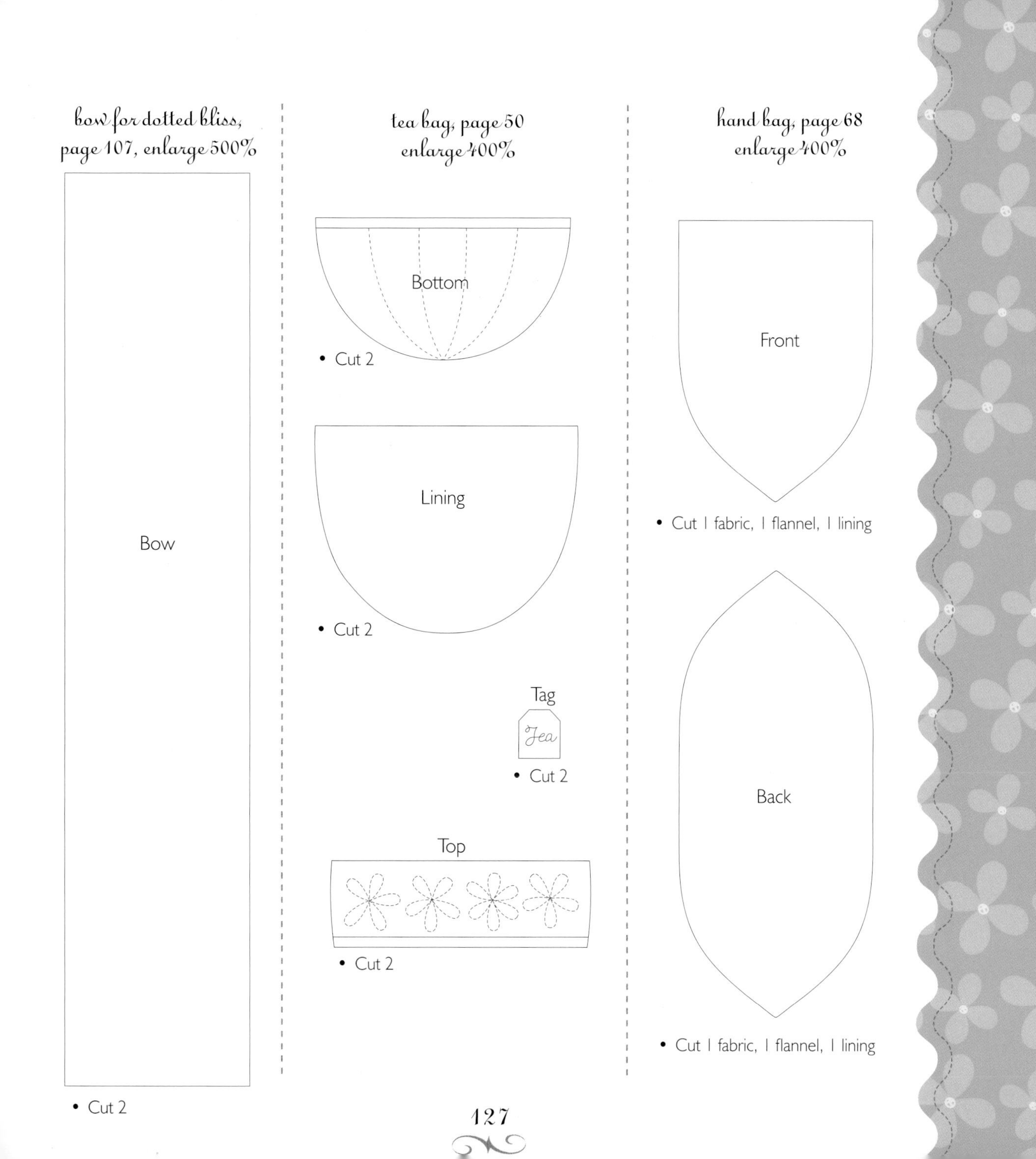

eco chic, page 70
enlarge 400%

checkmate, page 114
enlarge 400%

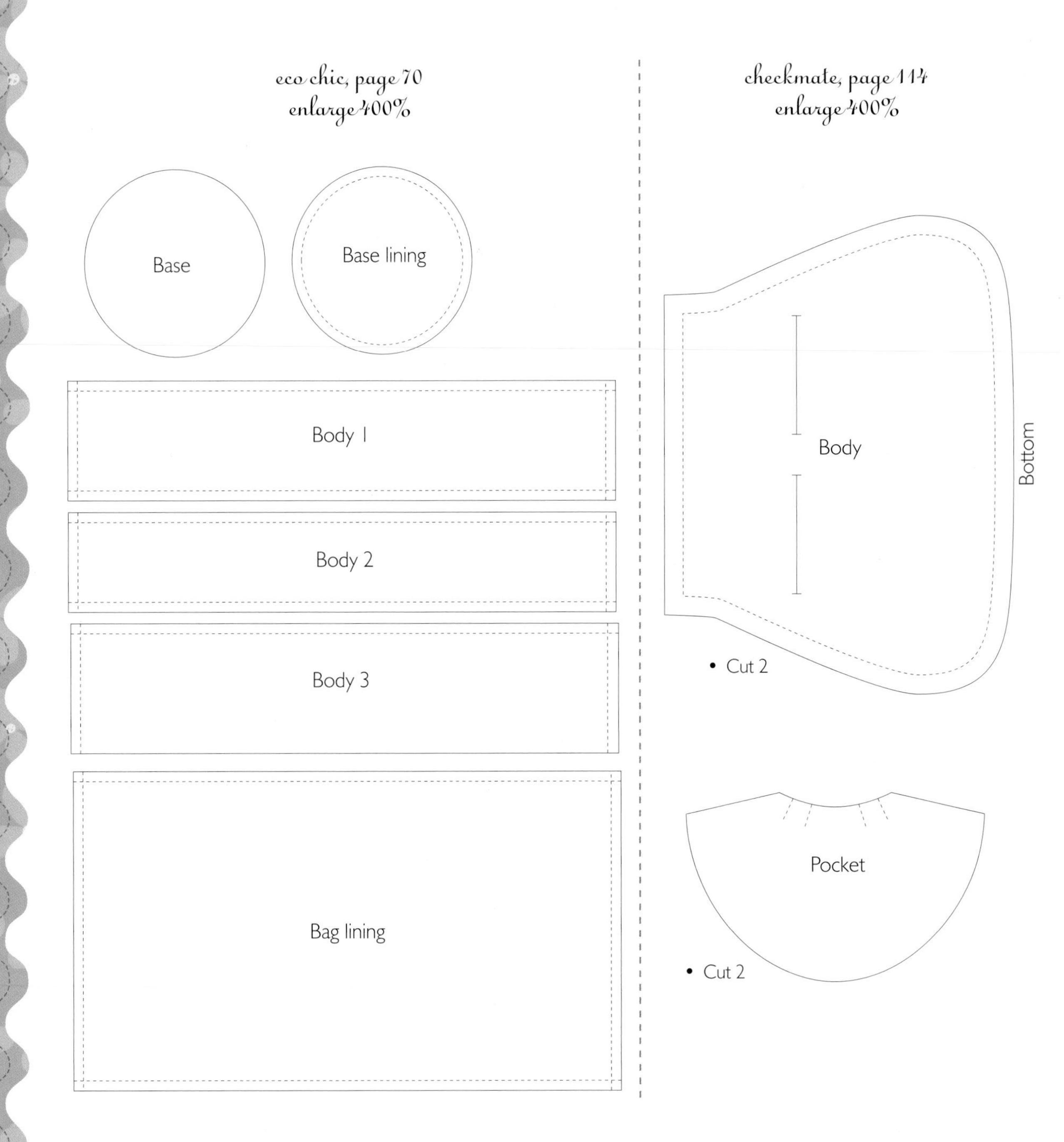

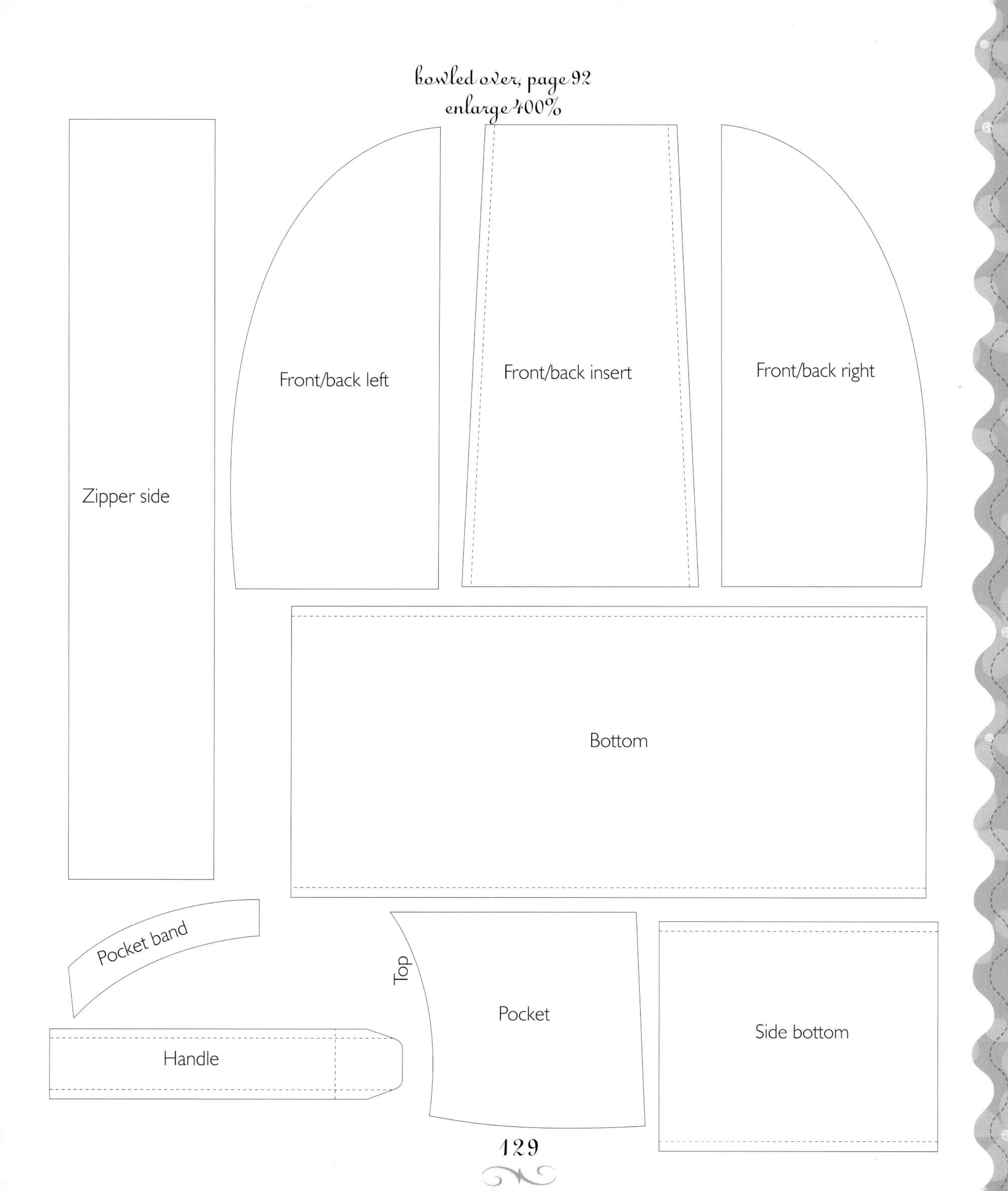
bowled over, page 92
enlarge 400%
Zipper side
Front/back left
Front/back insert
Front/back right
Bottom
Pocket band
Top
Pocket
Side bottom
Handle

about the designers

DORIE BLAISDELL SCHWARZ is living proof that blonde girls from the Jersey shore aren't all about manicures and ultratight jeans. Dorie currently lives in a small town in Illinois with her husband and their young daughter. When she's not sewing or crafting, she keeps a craft blog at tumbling-blocks.net/blog/.

RACHEL FIELDS has enjoyed bowling for many years, most currently on the Nintendo Wii. Although she can never seem to score above 100 points per game, she still maintains a deep love for the fashion of the lanes, which is where she drew her inspiration for designing the Bowled Over bag in this book.

NADJA GIROD started sewing on her grandma's sewing machine when she was 12 years old. After studying fashion design in Berlin and working in Sweden and Spain, she started her own accessories label. Take a look at www.smil.biz to find out more.

WENDI GRATZ lives with her family and her sewing machine in western North Carolina. In high school, she skipped home economics in favor of wood and metal shop. Now she makes fun clothes, funky dolls, and all kinds of quilts. You can see her work at www.wendigratz.com.

CASSI GRIFFIN resides in the mountains of Idaho with her three children and more than a dozen pets. She has contributed to numerous craft books and national magazines, and writes a craft blog called Bella Dia (http://belladia.typepad.com), where you can check out her newest projects, tutorials, and latest thrift finds.

AUTUM HALL lives in North Carolina with her husband, two children, and too many pets to name. Her work is featured in other books in this series, including *Pretty Little Potholders* (2008), and *Pretty Little Patchwork* (2008).

REBEKA LAMBERT is a self-proclaimed fabric addict. After taking a break during the early years of motherhood, Rebeka has returned to crafting full force. You can catch a glimpse of her life on her craft blog, http://artsycraftybabe.typepad.com, and see her latest creations at www.artsycraftybabe.etsy.com.

MICHA MAE MELANCON was born with a cool blue eye and a sense for excitement and danger in southern Louisiana. Micha's artistic influences include accordion music, Godzilla, New Orleans jazz babies, cakewalks and bake sales, white alligators, and recycled bridesmaid's gowns circa 1970. Visit her online store at www.bakingwithmedusa.etsy.com.

On most days, you'll find designer MORGAN MOORE whipping up delicious treats in her kitchen or designing new products for her online shop—all while chasing around a two-year-old toddler. Visit her on her blog, One More Moore, at morganmoore.typepad.com.

NATHALIE MORNU'S very first sewing project—a purse assigned in a 7th-grade home economics class—languished for years, unfinished, in the back of a closet. Later, she learned to sew garments with her aunt. Nathalie is the co-author of *Survival Sewing* (2007) and the author of *Cutting-Edge Decoupage* (2007) and *A is for Apron* (2008).

JOAN K. MORRIS has contributed projects to numerous Lark books, including *Pretty Little Potholders* (2008), *Pretty Little Pincushions* (2007), *Cutting Edge Decoupage* (2007), *Creative Stitching on Paper* (2006), *Exquisite Embellishments for Your Clothes* (2006), *Beautiful Ribbon Crafts* (2003), *Gifts for Baby* (2004), *Hardware Style* (2004), and *Hip Handbags* (2005).

AIMEE RAY has been making things for as long as she can remember. She works as a graphic designer in the greeting card and comic book industries, and is the author of *Doodle Stitching* (2007), a book of contemporary embroidery designs and projects. See more at www.dreamfollow.com.

According to family legend, ELIZABETH SEARLE began sewing in her crib. She's a self-taught fabric artist and a creative sewing teacher for various groups. Elizabeth has worked as a dressmaker for many years, and has contributed to several Lark books. She is the author of *Fun-to-Wear Fabric Flowers* (2006).

VALERIE SHRADER made a pair of pink culottes when she was 11 and has loved fabric ever since. She's on the staff of Lark Books, and has written and edited many books related to textiles and needlework. Recently, she celebrated her midlife crisis by purchasing three sewing machines in one year.

LAURRAINE YUYAMA used to work as a custom picture framer, but decided to start creating her own art instead. When she's not crafting, she spends time with her little girl and husband. Her work is sold internationally from her home-based studio in Vancouver, Canada, and through her online shop, www.patchworkpottery.com.

acknowledgments

A big cheer to all the designers who created such gorgeous, original works. We simply couldn't have done this without you! Thanks for sharing your brilliant talent and keeping us all inspired.

Cassie Moore deserves much praise for helping round up a terrific collection of the sweetest bags ever. Thanks also to Susan Brill, Dawn Dillingham, and Kathleen McCafferty, the stellar editorial team who worked on this book. The art production team of Will Ketcham and Nicole Minkin kept the book running smoothly on track.

Susan McBride's charming illustrations add whimsy to every page, and Orrin Lundgren's spot-on templates make the projects a breeze to follow. Photographer Stewart O'Shields helped the projects shine under the bright lights, with the support of his multi-talented assistant, Megan Cox, who somehow found time to pose with some of our bags while juggling a thousand other duties. The lovely Maggie West also modeled for us.

Finally, many, many thanks to art director Megan Kirby for presenting each purse with such personality.

index